EXPLODING STEAMBOATS, SENATE DEBATES, AND TECHNICAL REPORTS

The Convergence of Technology, Politics, and Rhetoric in the Steamboat Bill of 1838

by
Professor R. John Brockmann, TSSF
University of Delaware

Technical Communications Series
Series Editor: Charles H. Sides

Routledge
Taylor & Francis Group

LONDON AND NEW YORK

First published 2002 by Baywood Publishing Company, Inc.

Published 2018 by Routledge
2 Park Square, Milton Park, Abingdon, Oxon OX14 4RN
52 Vanderbilt Avenue, New York, NY 10017

First issued in paperback 2018

Routledge is an imprint of the Taylor & Francis Group, an informa business

Library of Congress Catalog Number: 2001043304

Library of Congress Cataloging-in-Publication Data

Brockmann, R. John.
    Exploding steamboats, Senate debates, and technical reports : the convergence of technology, politics, and rhetoric in the Steamboat Bill of 1838 / R. John Brockmann.
        p. cm. -- (Technical communications series)
    Includes bibliographical references and index.
    ISBN 0-89503-266-X (cloth)
        1. Steamboats--Law and legislation--United States--History. 2. Steamboats--Safety regulations--United States--History. I. Title. II. Series.

KF2550.B76 2002
343.7309'65--dc21

                                                                            2001043304

ISBN 13: 978-0-415-40411-2 (pbk)
ISBN 13: 978-0-89503-266-9 (hbk)

Table of Contents

LIST OF FIGURES

TABLES

Acknowledgments

Thanks to John Miller, my student researcher, and friend; the staff at the U of D Interlibrary loan that put up with me and all my requests; Jack Shagna, a fellow researcher of steamboats; Peter Dorsey of St. Mary's College who showed me how some of the explosions were captured in a number of then-popular, now forgotten antebellum novels; and the Historical Society for Cecil County Maryland for allowing me to use the jigsaw puzzle, "Blown Up Steamboat," as my cover image.

Dedication

Soon conversation chang'd to play
Upon topics of the day;
News, stale enough in distant town,
Just in the "Hollow" ushered down;
Murder and rapine, loss by fire,
Steamboat explosions, extra dire;—
Till last at politics they went,
And much of breath and speech were spent . . .

Josiah Dean Canning, "A Night in a Country Inn"
in *Conneticut River Reeds.*
Boston: J. Cupples, 1892: Ins. 257-264.

Introduction—
The X on the Draft Bill

A sloppy **X** (see Figure 1) removed lines 7 to 12 in the seven-page S.1. draft bill considered by the Senate in January 1838. To be exact, the **X** occurred in the seventh section of the first bill considered in that second session of Congress, a bill tentatively entitled

> To provide for the better security of the lives of passengers on board of vessels propelled in whole or in part by steam.

And, dates on the document suggest that the **X** was scrawled during a meeting of the Select Senate Committee chaired by the Honorable Felix Grundy of Tennessee.

The way Senator Grundy handled this bill and worked his Senate Select Committee in 1837 and 1838 was "fast-tracking" indeed. Only twenty-four hours elapsed from the time President Van Buren re-introduced the topic of safety aboard steam vessels in his State of the Union message on Tuesday December 5th to the moment Grundy was able to both frame its legislative language and get the Senate to create a Select Committee to consider the bill on Wednesday December 6th. Six weeks later, someone in the Select Committee put the **X** on the page of draft legislation, and then the Committee sent the approved, amended bill to the Senate for a vote on Wednesday January 24th.

In this simple **X**, many antebellum worlds converged. In the scrawled **X** was the world of steamboat technology in its earliest decades; the first steamboat on the Western waters, Shreve's *Washington*, had only made its maiden voyage in 1816, and, in June that same year, became the first steamboat in the West to have its boiler explode [1, p. 359]. Within three decades, the muddy water of the rivers, the nonstandard strengths of the brittle cast iron, the poorly trained engineers, and the mistaken understanding of steam power combined to create a lethal mixture that killed nearly 3000 people.

In the **X** there was certainly the public hysteria that arose from the hundreds of lives lost in the explosions of steamboat boilers. These deaths gripped the public attention for over a decade because Americans were caught in a paradoxical

1

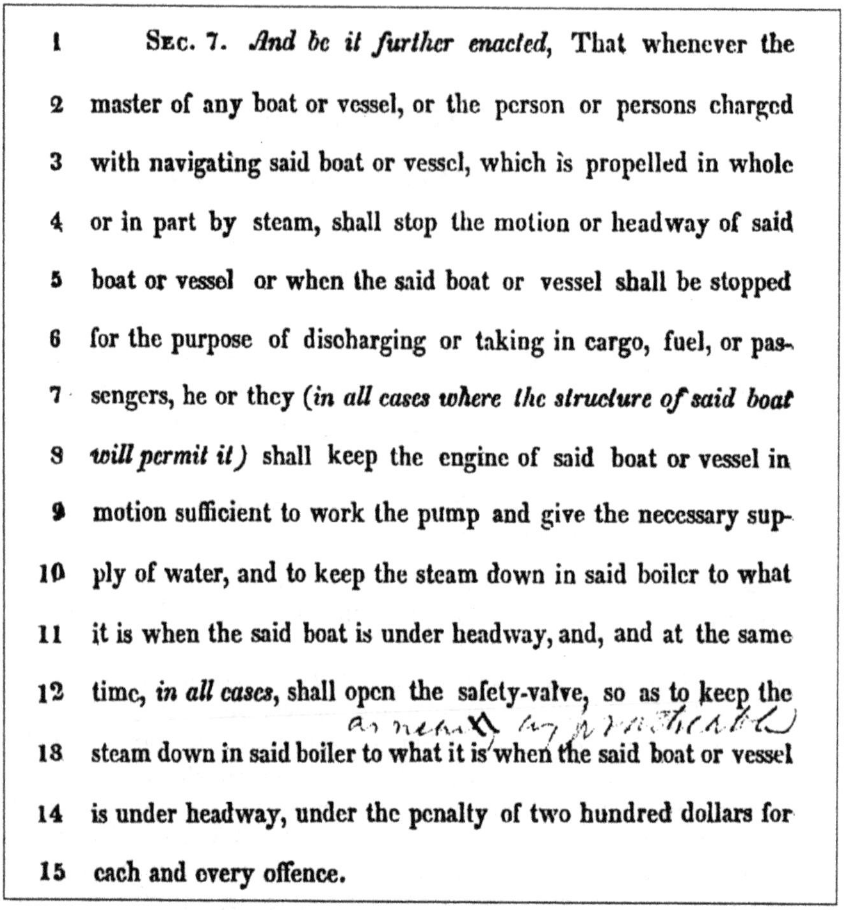

1   SEC. 7. *And be it further enacted,* That whenever the

2   master of any boat or vessel, or the person or persons charged

3   with navigating said boat or vessel, which is propelled in whole

4   or in part by steam, shall stop the motion or headway of said

5   boat or vessel or when the said boat or vessel shall be stopped

6   for the purpose of discharging or taking in cargo, fuel, or pas-

7   sengers, he or they (*in all cases where the structure of said boat*

8   *will permit it*) shall keep the engine of said boat or vessel in

9   motion sufficient to work the pump and give the necessary sup-

10   ply of water, and to keep the steam down in said boiler to what

11   it is when the said boat is under headway, and, and at the same

12   time, *in all cases,* shall open the safety-valve, so as to keep the

13   steam down in said boiler to what it is when the said boat or vessel

14   is under headway, under the penalty of two hundred dollars for

15   each and every offence.

Figure 1. The X in Lines 7 to 12 in Section 7 from draft Bill S.1.,
25th Congress, Second Session, January 9th 1838.

feeling that steamboats were simultaneously one of the first technological break-
throughs of the 19th century—a "gift from God" [2; 3, p. 212]—yet they were also
instruments of unprecedented destruction and death. On the very eve of the debate
on Senate bill, the *Charleston Mercury* spoke of this:

> We suffer the mighty despotism of steam to roll over us with the cold and
> grinding regularity of fate, and, shutting our ears to the shrieks of its victims,
> congratulate ourselves that on the whole we are more powerful, rich, and
> civilized that could have been without it. The community are [sic] responsible
> for the use they make of this power, so vast both for good and evil . . . [4, p. 2].

And, far to the west and north, the fledgling *Chicago American* chimed in:

> Here is another horrid list to be added to the sacrifices of human life, which are
> not almost constantly occurring on our steamboats. Can or will nothing be
> done to stay an evil whose frequency and devastation are making it as a
> pestilence among us? [5, p. 3].

The paradox of steamship technology reached deep into the American soul, surfacing for decades in the popular culture of the day—parlor songs, stories, and folklore. Moreover, the explosions and deaths were kept in the forefront of the American imagination because they were repeatedly splashed across the front pages of antebellum America's newspapers.

In the world of national politics, attempts to ensure safety aboard steamboats rose and fell with the hysteria and eventually came to involve many of the leading politicians of the era—Presidents Andrew Jackson and Martin Van Buren and Senators Daniel Webster of Massachusetts, John C. Calhoun of South Carolina, Felix Grundy of Tennessee, and Thomas Benton of Missouri.

Finally, in the X was the world of technological persuasion used by a group of nationally reputed scientists at the Franklin Institute in Philadelphia when they produced the first federally funded report to Congress focusing upon a technological catastrophe. The twentieth century reports on the Three Mile Island catastrophe and the Challenger O-Ring explosion all find their fountainhead in this collaborative report managed by the grandson of Benjamin Franklin, Alexander Dallas Bache.

Bache's *General Report* was characterized by one eminent historian of antebellum technology, Professor Bruce Sinclair, in the following way

> There was no question of the report's theoretical soundness, just as its
> practical value was obvious. It was also cast in near-perfect form to secure
> legislative actions [6, p. 19].

However, no one has ever pursued the trail of this "near-perfect" report once it left the mechanic's institute in Philadelphia and was printed and distributed in Washington D.C. to senators and congressmen. No one has ever asked if the *General Report* helped to secure legislative action, or had its message lost in the hysteria surrounding the steamboat explosions such as described in the following letter to the editor during the considerations of the Bill:

> For some years past our feelings and sympathies have been almost daily
> wrought upon by the recital of the most shocking and heart-rending accounts
> of the destruction of human life by the explosion of steam-boilers, and lately
> these shocking occurrences appear more frequent. If it is in the power of
> human ingenuity to prevent it, no effort or expense should be spared to effect
> this most desirable object. The ease and advantages arising from conveyance
> by steam make it of the first importance that it should be rendered safe. By this
> kind of conveyance the legislative bodies of our country, and our wives and
> children, are daily conveyed from one section of the country to another, and

from the present state of things, one's mind is in a continual state of distressing doubt whether they will ever meet the friend, the wife, or child, that they part with on board of a steamboat [7, p. 3].

If anyone ever did ask about the success of this report, they would have found that the Steamboat Law of 1838 finally passed by Congress and signed into law by President Martin Van Buren did not work.

The deaths, explosions, and shipwrecks continued; more explosions occurred after the law was passed than had occurred in all the time from 1824 to 1838 [8, pp. 98-99]. As *Report 241* in the first session of the Congress following passage of the law noted:

The act of Congress referred to, has undoubtedly contributed in some degree to the public security; but we have abundant proof that it falls far short of effectually shielding the public from those disasters which prompted its adoption. Within the last year (1838-1839) about 200 lives have been lost by the causes complained of; exceeding the average of former years [9, p. 2].

An 1840 communication from some three dozen proprietors and managers of steam vessels concurred when they noted:

Your memorials believe that few opinions are more erroneous than that which ascribes to the provisions of the existing law a generally increased safety for persons and property carried in steamboats. This may appear from the many accidents or disasters of a serious character which have taken place during the short period in which this law has been in force. The number of these accidents on the western waters during the last year is stated to have been forty; which may serve to convince Congress that the appropriate remedies for these disasters are not furnished by this law [10, p. 4].

And finally, a report created in 1850 by a Committee of the Citizens of Cleveland in Relation to Steamboat Disasters also concurred when it noted:

And further, it is our opinion that the inspection, provided by the law of Congress, has been inefficiently executed, and that the law itself is radically defective [11, p. 7].

The best information from the most capable scientists and technologists of the antebellum era argued against what the **X** did. The best minds in Congress should have read and heeded their warning. And yet the **X** was struck, the law was erroneously passed, and the deaths and explosions continued. Why?

Endnotes

1. Horrid Accident, *Weekly Recorder*, Chillicothe, Ohio, June 13, 1816.
2. The Rev. James T. Austin in 1839 [quoted in 2].
3. Hugo A. Meier, Technology and Democracy, in *Technology and Change*, John G. Burke and Marshall C. Eakin, Editors, Boyd and Fraser, San Francisco, 1979.
4. *Charleston Mercury*, 26 (4328), November 6, 1837.

5. *Chicago American,* August 26, 1837.

6. Bruce Sinclair, *Early Research at the Franklin Institute: The Investigation into the Causes of Steam Boiler Explosions, 1830–1837,* The Franklin Institute, Philadelphia, Pennsylvania, 1966.

7. Letter to the Editor, *National Intelligencer, 26*:7923, July 6, 1838.

8. David John Denault, *An Economic Analysis of Steam Boiler Explosions in the Nineteenth-Century United States,* Ph.D. Dissertation, University of Connecticut, 1993. (Available from UMI Dissertation Services, Ann Arbor, MI.)

9. Senate # 241, 25th Congress, 1st Session.

10. *Memorial of Sundry Proprietors and Managers of American Steam Vessels on the Impolicy and Injustice of Certain Enactments Contained in the Law Relating to Steamboats,* New York, p. 4, 1840.

11. *Proceedings of a Meeting and Report of a Committee of the Citizens of Cleveland in Relation to Steamboat Disasters on the Western Lakes,* Steam Press of Harris, Fairbanks & Co., Cleveland, Ohio, p. 7, 1850. See also p. 16: "The law now in existence, passed July 7, 1838, entitled 'An Act to provide for the better security of the lives of passengers on board of vessels propelled in whole or in part by steam,' has many valuable provisions in it, but it is defective in several essentials."

Figure 2. "Lost on the Steamer *Stonewall* or, Mamma! Why
Don't Papa Come Home?" [6]

Listen, darling to a tale of woe;
While on her downward way,
A boat caught fire, no human pow'r the raging flames could stay;
The *Stonewall* on that night was doom'd!
No help was nigh to save!
Alas, my child, but very few escaped a wat'ry grave.
Alas, my child, but very few escaped a wat'ry grave.
But now, Mamma!
Dry up your tears,
Cease, cease sighs;
For by and by we'll meet Papa,
Far up above the skies.

CHAPTER 1

Steamboat Politics and Steamboat Society

The Senate Select Committee chaired by Senator Felix Grundy of Tennessee in January 1838 was comprised of some of the most powerful men in the Senate. Three had been or would soon be candidates for President or Vice President of the Republic—Daniel Webster of Massachusetts, John C. Calhoun of South Carolina, and Thomas Hart Benton of Missouri. These three were joined by some lesser known senators including Garret Wall of New Jersey; Thomas Clayton, former Chief Justice of Delaware; and Robert John Walker from Mississippi who would later be elected governor of the new state of Kansas. For all these senators, December 1837 and January 1838 was a critical time in national politics:

- there was still the national recovery to create after the banking panic of 1837, and Congress was reconsidering the entire banking system, the relationship of the Treasury and banks, and the national currency [1, Chapters 3–5];
- there was a revolt in British Canada called the *Patriot War*, and, in the confusion, an American steamship, the *Caroline* (see Figure 3), was commandeered, burned, and sent plunging over Niagara Falls [1, Chapter 8];
- there was fighting continuing in Florida under the leadership of Gen. Zachary Taylor in the long war with Osceola's Seminoles [1, Chapter 8; 2, p. 260] and volunteer troops from Senator Benton's home district were being dispatched to fight [3, p. 288];
- and, most important of all, the problem of slavery had seized the focus of many in Congress as new territories in the west such as Texas sought admission into the United States or appeared soon to seek admission. In this bitter quarrel a month earlier, an abolitionist had been lynched, and, in the national reaction, Senator Benjamin Swift from Vermont rose in the Senate to suggest prohibiting further annexations as well as abolishing slavery in the areas already in the Union, e.g., the District of Columbia. Senator Calhoun of South Carolina counterattacked in a series of blistering, lengthy speeches defending slavery and keeping the Senate in debate through the holidays until January 13, 1838 [1, p. 103].

Figure 3. Steamer *Caroline* plunging over Niagara Falls, December 1837.
(Reprinted with permission from the Canadian Heritage Gallery.)

In addition to all of these items gripping the Congress at large, there were a number of personal problems drawing the attention of the members of this Senate Select Committee who were to consider the steamboat legislation. Daniel Webster was preparing for another presidential run in the Whig Party and did not even reach Washington City for the Fall Session until December 29th 1837. His political ascendancy contrasted with the precipitous fall in political favor experienced by Felix Grundy, Chair of the Committee.

Grundy had seen the reactions from his Tennessee constituents to his vote earlier that Fall on the banking proposition in which he had supported the Administration with the now obvious result that Grundy was not going to be returned to the Senate after the next election. Grundy wrote to his old friend and former President, Andrew Jackson:

> My present impression is that I shall retire from public life, and endeavor to repair my fortune, which has been shattered by my public service, for which I have received but poor compensation from the people whose interests I have endeavored to promote. This, however, is my consolation, I know I have served the people faithfully, if they do not know it—and the time will come when they know it too [4; 5, p. 317].

Finally, Committee member Thomas Benton probably felt he needed to leave Washington City and return as soon as possible to his home in St. Louis to mourn the deaths of his mother, his mentor, and Col. Gentry, leader of the Missouri volunteers in the Seminole War who had just been killed at the battle of Okeechobee [3, p. 288].

With so much going on in Congress, as well as in the individual lives of the senators, why do steam boilers and steamboats involve so many and move so swiftly in Congress? Understanding the reason behind the Twenty-fifth Congress's Bill S.1. requires one to look back some fourteen years to the Eighteenth Congress and its first concerns about steamboat safety.

New York Harbor, May 15, 1824, 7:00 PM

How sober and discrete ought they to be who have charge of machinery capable of accomplishing such terrible mischief in a moment! [7, p. 239].

Everything had quickly changed after Lawyer Daniel Webster had persuaded the Supreme Court to break up the Fulton-Livingston monopoly on steam traffic on the Hudson (see Figure 4) [9]. The Fulton-Livingston monopoly allowed Fulton's **low**-pressure steamboats to control the large New York City traffic, but now **high**-pressure boats such as the *Aetna* could legally be moved by its owners from the Delaware River into the more lucrative Hudson River traffic (see Figure 5). These **high**-pressure boats were faster and cheaper to run, but their arrival precipitated a savage commercial war on the Hudson River: rates were ruthlessly cut, collisions, some intentional, increased, and "accidents" became more frequent [10, p. 204]. In this atmosphere of commercial warfare, on the evening of May 15th, a steam boiler blew in one of the **high**-pressure boats challenging the Fulton-Livingston monopoly (see Figure 5).

A report on this was made in a letter from a passenger to friends in Philadelphia:

It is with pain I inform you of an awful occurrence that took place at 7 o'clock, last evening on board the steamboat, *Aetna*, Captain Thomas Robinson,—when, about seven miles from, and in sight of this city, her boilers bursting with a noise like thunder, and throwing the pieces upon the quarter deck, where I had the minute before been standing. I had walked to the bows when the explosion took place; and thanks be to the Almighty that I am one of the few that escaped unhurt. O! the awfulness of the scene! [11, pp. 140-142; 12, pp. 190-191].

Out of a crew and passengers of 34, 12 died and 9 were wounded. A week later the newspaper *The Niles Register* noted:

If we recollect rightly, these (the *Aetna* and the *Eagle*) [another high-pressure steam boat that exploded at the time], are the first fatal accidents that have

Figure 4. View of New York Harbor—Steamboat on left at
mouth of Hudson [8, p. 436].

happened in steam boats, either in the waters of the Chesapeake or those of
New York and the parts adjacent [14, p. 192].

The *Aetna* explosion disturbed the public to such an extent that the Reverend John
Stanford printed up a sermon since he esteemed,

> it a duty devolving on me, to draw the general outlines of this calamity,
> and communicate to you some of those solemn and interesting sentiments
> expressed by the immediate sufferers in their dying hours; in the devout hope
> that they may convince you of the frailty of life, and the imperious necessity
> of being prepared to meet your God [15].

Four Days Later—Washington City, May 19, 1824

Representative Samuel Vinton (Ohio) stood in the House to declare that "A
country agitated with terror and dismay looks to us for protection" [16, p. 2671; 17,
p. 27]. He urged an immediate ban on all high-pressure steam engines "to give
practical security to the enjoyment of this most valuable invention [steamboats]."

Three days later, on Saturday May 22nd, in a House where "the sensation
produced by the news had not yet subsided in every part of the Hall" [14, p. 200;
18, pp. 2670–2671], a draft bill was reported out from the Commerce Committee.

For PHILADELPHIA—24 miles land carriage. *Through in Less than One Day—Fare $4.*

CITIZEN'S LINE, connected by Steam Boat Ætna, Captain Robinson, starting every week day, at 6. A. M. from this city to the village of Washington; thence by land, over a new, smooth and level road, to Bordentown; from there, by Steam Boat PENNSYLVANIA, Captain Kellum, to Philadelphia—Breakfasting on board of one boat, and dining on board of the other.— Private parties, and families, will find this a desirable route, as it prevents the necessity of lodging on the road, and avoids travelling by night.

The coaches and stock of horses are in fine order, with careful and attentive drivers, and the two steam boats, with the politeness and attention of the captains, attached to the line, are not surpassed.

The above advantages combined, the Proprietors can with confidence recommend this route to their friends and the public, for expedition, comfort, and variety—far superior to any route between the two cities. For seats, apply to THOS. WHITFIELD, at the old mail and general coach office, No. 1 Courtlandt street, and on board of the Ætna, foot of Liberty st. N. River.

U. S. Mail Coach, daily at 2 P. M. arriving in Philadelphia next morning at 6 A. M.

Extra coaches with two or four horses, and expresses, at any hour.

LYON, WARD & BAYLEY,

n 24   d2w   Proprietors.

Figure 5. *Aetna* ad in *New York Evening Post* the day after the explosion, May 1824 [13, p. 1].

The bill shifted from the outright abolition of all high-pressure steam engines and called instead for boiler inspections and licensing, as well as the implementation of safety valves on all steam boilers—including a locked valve inaccessible to the boiler engineer who might tamper with it. Finally, the draft bill provided a $500 fine to punish any tampering with the safety valves. This fine was intended to solve the problem of "racing." Racing required increased speed which could only be had by tampering with the safety valve to cause the boiler to hold additional steam . . . and increasing the danger of exceeding the capacity of the boiler to contain the steam:

> Boats are constructed upon the cheapest plan, and with a view almost exclusively to velocity of motion. The traveler is always in haste to go forward, and hence, the fastest running boats obtain the highest character. This naturally keeps up a constant and most dangerous competition, that has been, and ever must be fruitful in disaster [18, p. 2672].

When the *Moselle* exploded in 1838, an editorial in one of the town's papers made this observation about the public fascination with speed:

> For this result we are in part to blame; we plead guilty, in common with other presses, of having praised the speed and power of the boat—a circumstance which doubtless contributed to inflate the ambition of the captain and owners to excel others in rapidity. If the public are to have any security against steamboat accidents, the press must change its tone. Boats must be praised for their comfort, convenience and the care and discretion of their commanders—but not for their speed [19, p. 342].

Vinton followed up this 1824 draft bill by putting into the House record a ten-page report, *Report 125*. *Report 125* described "racing" in the following manner:

> From habitual impunity the engine workers disregard the danger (of bursting boilers from having overloaded safety valves) and rather than suffer a boat to pass them, will increase the pressure of the steam to a dangerous extent [20, p. 2].

Report 69, issued a few months later, also referred to this problem:

> Engineers and even captains of steam boats, have a great propensity to overload their safety valves. It enables them to make a show of great speed at leaving a port, which gratifies their vanity [21, p. 16].

On Monday May 24th, nine days after the explosion of the *Aetna* in New York harbor, the House of Representatives took up the bill for a vote but seemed wary of passing it. There were repeated complaints by members who spoke of insufficient time in the session to consider the bill, and that some reports supporting the Bill that had been entered into the record, such as the one from Philadelphia (*Report 125*), were seven years old [22, 23], and so members asked for additional information concerning the boiler materials and the problem of racing. The result was that the House voted 67 to 47 to postpone action on the bill, and subsequently, on Wednesday 26th, the Secretary of the Treasury was instructed to provide the additional information by investigating the matter and reporting back to the House:

> **Resolved**, That the Secretary of the Treasury be instructed to inquire and report to this House, at the commencement of the next session of Congress, what are the material causes of those fatal disasters which have so frequently occurred on board steam boats in the waters of the United States . . .

William Crawford, Secretary of the Treasury, did indeed report back on the 31st of January the following year with a lengthy report (*Report 69*) [21] which he prefaced in such a way as to undermine any legislation being drawn from the pages of the report:

> I am of opinion, that legislative enactments are calculated to do mischief, rather than prevent it, except such as subject the owners and managers of those boats to suitable penalties in case of disasters, which cannot fail to render the masters and engineers more attentive, and the owners more particular in the selection of those officers.

Both *Report 125* and *Report 69* were collections of unorganized disparate materials and, which, in the case of *Report 69*, included correspondence from, among others, the British House of Commons, steamboat masters, and Treasury collectors in various ports. Neither report's findings were cohesive and, at times, they were quite contradictory.

In the end, the Eighteenth Congress did not pass the bill. The pressure on Congress to do something about the steamboat explosions was matched by a lack of understanding regarding the technology and of consensus about what exactly to do. It was also unclear where to lay the blame:

- if the problem lay with technology, then the following elements of the bill proposed by Vinton would take care of that: each boiler would be inspected and its safe working pressure and general condition would be examined; to have two safety gauges, to ensure that the gauges were only loaded to one-third or one-sixth of the pressure that could be sustained by the boiler, and this would be applicable to both low-pressure and high-pressure engines. This approach was directed at controlling the dangerous effects of technology.
- if the problem lay with negligent operations by the engineer, then the following element of the bill proposed by Vinton would take care of that: one of the two safety gauges would be in the control of the engineer while the other would be in the control of the captain; moreover, any evidence of tampering would be grounds for a $500 fine. This approach was directed at humans being able to make clear decisions in a situation where control of the technology was possible—the fine would simply contain and limit bad decisions.

Vinton's bill took the approach of attacking both problems, yet Secretary of Treasury Crawford in *Report 69* tried to dismiss the technology problem and emphasize the negligence aspects with their civil liability solutions. In the end of the debate, the solutions to the explosions was more confused than when it had all begun.

Endnotes

1. Major L. Wilson, *The Presidency of Martin Van Buren*. University of Kansas Press, Lawrence, Kansas, 1984.
2. *The Papers of Daniel Webster, Correspondence, Volume 4: 1835-39*, Andrew J. King, Editor, University Press of New England, Hanover, New Hampshire, 1989. A lynching of an abolitionist by a mob in Alton, Illinois on November 7 is what touched off the controversy and was the proximate cause of the Vermont anti-slavery forces presenting these bills in the Senate.
3. William Nisbet Chambers, *Old Bullion Benton: Senator for the New West*. Little Brown, Boston, 1956.
4. *Correspondence of Andrew Jackson*, Edited by John S. Bassett, Carnegie Institute, Washington, D.C., 1931. Grundy to Jackson, 2/9/1838.
5. Joseph Howard Parks, *Felix Grundy: Champion of Democracy*, Louisiana State University Press, Baton Rouge, 1940.
6. Words by J. Murphy, Music by G. W. Brown, Balmer & Weber, St. Louis, Missouri, 1869.
7. *Niles National Register, 12,* 1817.
8. *The Mariner's Library*, C. Gaylord, Boston, 1834.
9. Maurice G. Baxter, *The Steamboat Monopoly: Gibbons V. Ogden, 1824*, Alfred Knopf, New York, 1972.
10. Dorothy Gregg, *The Exploitation of the Steamboat: The Case of Colonel John Stevens*, Ph.D. Dissertation in Political Science, Columbia University, New York, 1951.
11. S. A. Howland, *Steamboat Disasters and Railroad Accidents in the United States*, Dorr, Howland & Co, Worcester, 1846.
12. *Christian Journal and Literary Register, 8,* June 1824.
13. *New York Evening Post*, May 26, 1824 (No. 6831).
14. *Niles Register, 2*:11, May 22, 1824.
15. John Stanford, *A Discourse, Delivered in the New-York City Hospital on Lord's Day Morning, May 23, 1824*, K. Conrad, New York, 1824.
16. *Annals of Congress*, House of Representatives, 18th Congress, Session 1, 5/19/24.
17. John K. Brown, *Limbs on the Levee: Steamboat Explosions and the Origins of Federal Public Welfare Regulation, 1817–1852*, International Steamboat Society, Middlebourne, West Virginia, 1989.
18. *The Debates and Proceedings of the Congress of the United States, Eighteenth Congress—First Session*, Gales and Seaton, Washington, 1856.
19. Fred Erving Dayton and John Wolcott Adams (illustrator), *Steamboat Days*, Fredrick A. Stokes Co., New York, 1928.
20. *Report of the Committee on Commerce, Accompanied by a Bill for Regulating of Steam Boats, and for the Security of Passengers Therein*, 18th Congress, Session 1, House Reports, No. 125, Gales & Seaton, Washington, 1824 (Serial Set 106).

21. *Accidents on Board Steamboats* 18th Congress, Session 2, House Reports, No. 69, Gales & Seaton, Washington, 1825.
22. *Report of the Philadelphia Joint Committee Appointment by the Select and Common Councils on the Subject of Steam Boats*, Philadelphia, 1817.
23. *Niles National Register, 12,* 1817. "Such dreadful accidents may go so far to reduce the confidence of the people in these invaluable boats (under proper management) as to destroy a great part of their usefulness. Those who are conversant with the subject assert that such accidents always come out of carelessness. How sober and discreet ought they to be who have charge of machinery capable of accomplishing such terrible mischief in a moment!"

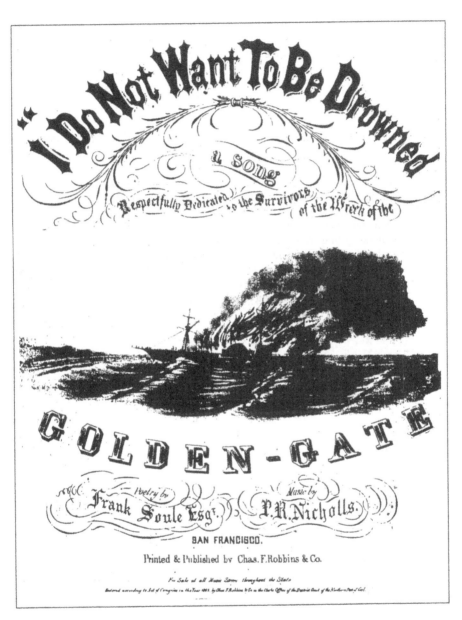

Figure 6. "I Do Not Want To Be Drowned" [1].

The following incident from which this Song [1] was written is related by Mr. A. Bates, one of the rescued passengers from the burning wreck of the late Steamship Golden Gate, destroyed by fire, near Manzanilla on the afternoon of July 27th 1862. Mr. Bates says: "While standing at the bows grasping a rope, a little girl, a lovely little child about 8 years of age, named Addie Manchester, came up to me, and said,

"O, mister, can you swim?"

"O, I told her I could, and would try to save her if she would do just as I told her. She said,

"Save me, do please. I don't want to be drowned!"

I showed her how to act—to get on my back and grasp me tightly, but that she must not choke me. She promised to do just as I told her. She was quite cool. Just as the fire got up to us, the steamer struck the bar. I jumped over, taking Addie with me. She held tight to me, and I struck out for the beach, not far off. The breakers were very high. I got past the first and second one in safety with my burden. After I passed the third one I found Addie was gone. I turned round and saw her going down behind me. A man on a plank who was passing, grasped her by the hair and pulled her on his plank. I saw she was safer than with me, so I continued on and was dragged on the beach. I lay on the beach insensible for about half an hour. When I came to, I saw Addie among the rescued passengers.

CHAPTER 2

Steamboat Technology

High-Pressure Steam Engines and Hulls that Ride on the Water

The steamboat united two cutting edge technological evolutions in the early 19th century, high-pressure steam engines and hulls designed to ride on the river water rather than cut through it as clipper ships were soon to do in the ocean [2, pp. 71-72].

The high-pressure steam engine was invented by Oliver Evans in Philadelphia around 1800 and had been described in his 1805 book, *The Abortion of the Young Steam Engineer's Guide* [2, pp. 50-58]. Evans advocated redesigning engines from producing the customary 2 to 4 pounds per square inch (psi) of pressure in the traditional Newcomen- and Watt-design engines to 100, 125, and higher psi [3, p. 47]. In the Newcomen- and Watt-design, steam produced power during its condensation phase when steam would shrink in mass and develop a vacuum that would **pull** a piston. These engines were termed "low pressure" or "atmospheric" engines, and, for their large size and heavy weight, could only develop low levels of power. Moreover, these "low-pressure" engines required a vast amount of fuel and cold water to produce this low amount of power. Evans's design, on the other hand, used the steam during its expansion phase to **push** a piston out, and to build up high pressure in the boiler to create more power using less fuel and requiring less weight in the engine. The shift to high-pressure engines was quickly made by United States steam engine designers between 1816 and 1830 [4, pp. 723-744; 5, p. 28].

However, from the first, "high-pressure" engines were resisted because "they were believed to be dangerous" [6, p. 281]. A biographer of Evans, their inventor, spoke of this early resistance:

> In the *Emporium of the Arts and Sciences*, Vol. 2, published in Carlisle, Pa., 1812, we find quite an extended account of the state of the steam engine at that period, and the feeling against the use of high-pressure steam is well illustrated by an account of the explosion of one of Trevethick's boilers with fatal effect. This fear of the power of high-pressure steam dated from the time of Watt, who thought Richard Trevethick ought to have been hanged for using

it, and was a potent factor in the opposition which Evans encountered in his efforts to introduce his engine [7, p. 13; 8, p. 189].

Charles Dickens on his 1840 tour of America written up in *American Notes* had this wry observation regarding how passengers felt aboard high-pressure steamboats:

> It always conveyed that kind of feeling to me which I should be likely to experience, I think, if I had lodgings on the first floor of a [gun] powder mill [9, p. 231].

An illustration of a high-pressure steam engine contemporary with the draft legislation—a generation after Evans's invention—can be found in Renwick's *Treatise on the Steam Engine* [10] and illustrated in Figure 7.

Notice the absence of dials and gauges to display and control the level of the water or the pressure of the steam. The only safety devices were stop-cocks (see Figure 8) and safety valves (see Figure 9). The two stop-cocks would directly indicate the level of steam or water in the boiler by allowing the engineer to observe whether water or steam issued from them when they were opened, e.g., if water issued, then the valve was below or at the water line; if only steam issued then the water was below that stop-cock. The safety valve operated in a similar direct way. A weight attached to a rod kept a valve shut until the pressure of the steam exceeded the weight on the rod, at which time the valve would open and the steam pressure above the weight on the rod would be expelled up the waste pipe.

The design of the steam boiler illustrated in Renwick's book was nearly identical to that as conceived and invented by Evans in 1805. The only variation would be that Evans's design had two safety valves rather than the typical single one as shown in Renwick's illustration. In contrast to the fairly stable design of the steam boiler which powered the engine, the model, proportions, and manner of construction of the steamboat hull, the second 19th-century cutting edge technology, had rapid, radical changes.

As in most technological endeavors, the hull designers of western river boats first tried to duplicate the longstanding designs of traditional ocean-going vessels that required a rigid keel to withstand the impact of ocean waves, to offer a secure footing for tall sail masts, and to enable a sailing vessel to maintain its position relative to the direction of the wind by keeping the boat from getting blown cross-ways (see Figure 10). The river steam boat, however, had few if any waves to contend with, no tall masts to hold up, and no problem with wind and sail cross-pressures. Moreover, the depth of the western rivers was quite shallow.

Thus the keel disappeared over time, and the bottom of the steamboat hull became flatter allowing the hull to go into shallower river water. The height in the rear of a typical sailing vessel moved forward, and the superstructure got taller [11, p. 66]. Steamboat designers also extended the length and breadth of the deck to increase cargo space. And finally, since there were few waves to contend with on the rivers, the freeboard—the distance from the waterline on the hull to the top of

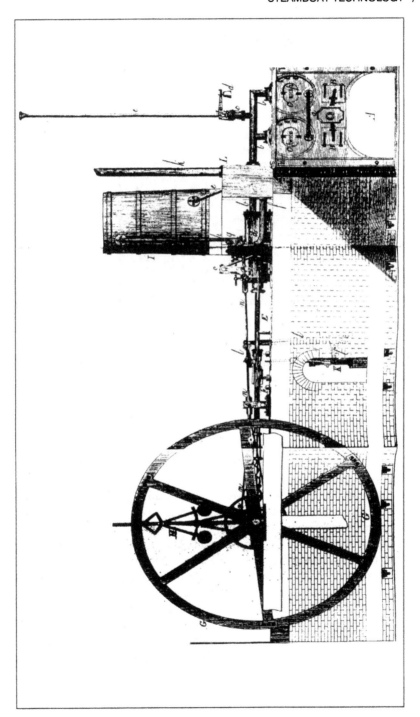

Figure 7. A high-pressure steam engine of the 1830s. a—steam issues out; b—safety valve; c—boilers.

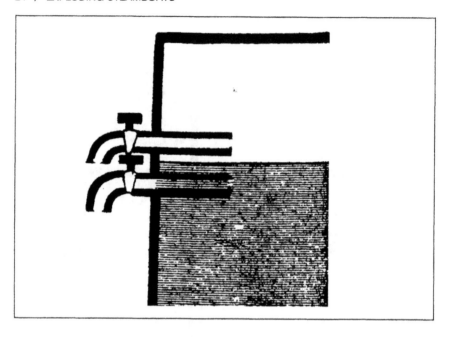

Figure 8. Two stop-cocks wherein if the bottom one were opened,
water should come out and thus the level of the water in the boiler would
be known, and if the top one were opened and steam issued that the
engineer would know that the right level of steam is occurring [10].

the deck—began to decrease until ships became nearly awash when fully loaded
(see Figure 11) [11, p. 79]. This flat hull design that increased length and width
also increased speed. To increase cargo capacity and minimize the depth of water
required to float, the weight of the vessels also steadily decreased so that the
planking and construction of the superstructure and decks was sufficient to hold
them together, but nothing more.

The result of all these changes to the steamboat design was that the lifespan of
a steamboat was much shorter than that of a typical sailing vessel at the time. The
average lifespan of a sailing vessel was 20 years: whaling ships ordinarily reached
40 years, and it was not uncommon for sailing ships to reach 60 years in service.
The average lifespan of a steamboat during the same period was five years [11,
p. 101]. When Tocqueville made his famous visit to the United States in early
1830s, leading to the creation of his book *Democracy in America*, he inquired
about the short life of American vessels. The answer quite nicely explains the
benefits of the short lifespan of a steamboat:

Q. It is said that your ships generally don't last very long?

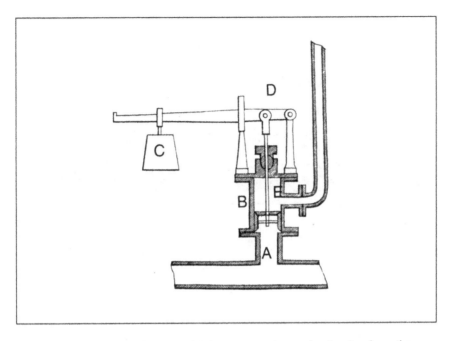

Figure 9. Pressure from the high-pressure steam pipe issuing from the boilers at (a) push up the sliding valve (b) which is kept in place by the external weight (c) through the rod at (d). When the pressure in the line rises above the weight of (c)—the weight of which should be equal to the maximum pressure of the boiler—then valve (b) is pushed up above the waste steam pipe at (e) and the extra steam maximum steam pressure is expelled [10].

Figure 10. An early 1824 steamboat hull design. Note the vestigial bowsprit and figurehead as on an ocean-going vessel.

Figure 11. A later 1837 steamboat hull design: Note that within 13 years, the bowsprit and figurehead are gone, and the deck is flatter and wider.

A. . . . What makes our ships last such a short time is the fact that our mer-chants often have little disposable capital at the beginning. It's a calculation on their part. Provided that the vessel last long enough to bring them in a certain sum, a surplus over the cost, the end is obtained. Besides there is a general feeling among us that prevents our aiming at the durable in anything; there reigns in America a popular and universal faith in the progress of the human mind. There are always expectations that improvements will be dis-covered in everything, and in fact are often right. For instance, a few years ago I asked the builders of steamboats for the North River why they made their vessels so fragile. They answered that, as it was, the boats would perhaps last too long because the art of steam navigation was making daily progress. As a matter of fact, the vessels, which steamed at 8 or 9 miles an hour, could no longer a short time afterwards sustain competition with others whose construction allowed them to make 12 to 15 [12, p. 645; 13, p. 121; 14, p. 356; 15].

It was this attitude of *daily* making progress that was, in fact, visually supported by the very rapid development of the structure of the steamboat hull. The problem was that the metal engines and boilers hidden away beyond the sight of most eyes did not experience such a rapid development for they were often salvaged and reused over and over again [11, p. 112]. William King noted that it was not uncommon for an engine to be salvaged and transferred four different times [16, pp. 182-183], and, in 1879, the *Times-Journal* in St. Louis published a list of 50 steamboats powered by salvaged engines [17].

The prolonged lives of the engines even made it into folklore in such stories as the following:

I think the sage old grandsires of the billiard saloon were fondest of recounting the odyssey of the engines of the packet *Joe Daviess*, once the proud property of the ubiquitous Harris family, who built her. Whey they sold her, she ran three trips and sank. The engines were dug up and placed in the first *Reindeer*. She made three trips and sank. Then the engines were dug up, resurrected and put into a second *Reindeer*. She made four trips and sank. Out of the wreck emerged the engines triumphantly and went into a new boat called the *Colonel Clay*. She made two trips and sank. Their next adventuring was in the new steamer *Monroe*. She ran all that season, and the best judges were satisfied that the hoodoo was broken. The next season she burned. The engines were now taken out and place in a mill at Elizabethtown, Pennsylvania, with a serene confidence that short of earthquake the mill could not sink, anyway. But it could burn, and did so the next year. The subsequent career of these machines is so far unrecorded [18, pp. 309-310].

Not only then was the engine not taking part in the art of steam navigation that was making *daily* progress, but because the engines and boilers were used over and over, they suffered. Transferring the boilers and engines meant that different crews would use them, and the varied rates and conditions of firing and use actually weakened the boilers.

Thus while both the engine and the hull structure of steamboats were among the most cutting edge technologies of the time, the public and Congress were misled into believing that the entire art of steam navigation was *daily* making progress with the engines and boilers "naturally evolving" into much safer machines. Rapid technological evolution was so taken for granted in the antebellum world that it became a major obstacle in the process of solving the problem of exploding steam boilers. More than once the following argument was used to stop legislative action from grappling with the steamboat problem:

It is better to leave the subject of the application of steam power to the propelling of boats to the sound discretion of those concerned, and to the improvements of the age, than to attempt, by any legislation of ours, to prescribe the particular kind of machinery to be employed [19, p. 1].

What Could Go Wrong with the Boiler Technology

Taking all this into consideration, it is a wonder that explosions have not been a more frequent occurrence. There was something radically wrong in the former construction of engines.
John Wallace, *The Practical Engineer* 1853 [20, p. 24]

There were a number of areas in which the boiler technology and operations were either inherently flawed or inherently vulnerable to operational errors. Moreover these design and operational flaws were complicated by engineers who had little or no training.

To begin with, it was difficult for cast iron manufacturers to create cylinders and rounded ends of uniform thickness and quality. Rarely did the cast iron from one manufacturer match the strength and quality of another, with the result that boilers performed unpredictably. Second, the interface of the boiler technology with the operators was quite limited and prone to mistakes. For example, the safety valve and the weight on the rod maintaining the level of maximum pressure (see Figure 9) were exposed to tampering on the part of engineers or firemen. If more power were needed for any reason, whether frivolous like racing or simply commanded by ill-trained captains who wanted to maintain a schedule, engineers or firemen could simply add more weight to the rod holding the valve closed or forcibly hold it closed (see Figure 12). Thus, it was easy for operators to exceed maximum design pressures in the boilers. (Moreover, even if the safety valves were not intentionally tampered with, simply the heat and water in the system tended to corrode the cast iron creating sticky valves that held pressures beyond the designed maximum [21].)

Third, the stop-cocks used to indicate the levels of steam and water (see Figure 8) and that seemed so foolproof and direct in their measurement were, in fact, prone to problems caused by foaming or by the tipping of the hull during loading and unloading. The foaming was created by the boiling of the water, and the foam could carry some water up to the stop-cock from several inches below—seeming to indicate that there was water at or above the stop-cock—even though it was several inches below.

Fourth, even if the cast iron had all been manufactured consistently, and even if the operator interfaces operated properly, the water in the western rivers was filled with silt. Thus, when the river water boiled off in the engines, the silt would accumulate on the bottom of the boiler and increase the possibility of having some parts of the boiler insulated by the mud taking in much less heat than those parts not having the silt insulation. This uneven heating could cause the boiler to lose strength and form. Uneven heating could also be caused by a tipped boat during unloading or loading procedures. One part of the metal of the boiler would be exposed to the heat more directly and would cause a flash of steam much faster than the rest of the boiler.

Fifth, and most insidious, the boiler and its steam powered both the engine propelling the boat as well as the pump bringing water into the boiler. Thus, in this double-duty situation, turning off the power to propel the boat would also disengage the pump and stop water from replacing that already evaporated into steam in the boiler. Theoretically, this should not have been a problem except that in the frequent short stop-and-go of steamboats coming in to pick up and deliver passengers or cargo, standard practice was to keep the fire stoked so that the vessel could quickly build-up propulsive power to pull away and continue on its journey. The safest procedure would have been to extinguish the boiler fire and rebuild it when it was time to get steam up, and, in fact, this procedure was called for a number times.

Among the papers of the 1837–38 Senate Select Committee in the National Archives today, there is a handwritten draft bill that had a section called

RESPECTFULLY DEDICATED TO MARINE ENGINEERS.

'THE ENGINE ROOM SONG.'

"OH THESE ARE THE DAYS OF QUICK DISPATCH!"

WORDS WRITTEN AND AIR ADAPTED

BY

CHARLES ROBERTSON.

AUTHOR OF "SAIL HO! HURRA!"

INVENTOR OF THE MAGNETIC DEVIATION DETECTER.

Figure 12. "The Engine Room Song"—
"These are the days of quick dispatch" [22].

O these are the days of quick dispatch.
When clearing thro' the deep.
We cheer on the work of our long watch,
While hardy messmates sleep.
We cheer on the work of our long watch,
While hardy messmates sleep.

Heave, heave on the coals,
Stoke well the fire.
Till steam begins to roar;
And show on the gauge all we require,
Then tend the furnace door.
We note on the gauge all we require,
Then tend the furnace door.
Now, now, start the gear,

She moves at last.
Now, now start the gear
Slow turns are giv'n to try:
"All right" says the chief: "Full speed" is
pass'd
The piston heaves a sigh.

"All right" says the chief: "Full speed" is
pass'd
The piston heaves a sigh.
Clink, clink rings the link, round goes the
crank
The valves how true they slide!
Each man to his place, his work and rank,
Bright engines are our pride.
Each man to his place, his work and rank,
Bright engines are our pride.

So now draw the fire, the steam blows off.
The engine's out of gear.
Then have a good wash, our duds we doff,
And in shore togs appear.
Then have a good wash, our duds we doff,
And in shore togs appear.

"Management of Engines and Boilers" in which the following was offered to prevent just this problem:

> On approaching and within one mile of each landing or stopping place, it shall be the duty of the fireman or person in charge of the Boiler to put the feed on the Boiler and on stopping the Engine to close the dampers in the chimney to stop the draught [thus effectively damping the fire under the boiler]. He shall then try the water gauge cocks after the Engine has stopped and if steam escapes from the lower water gauge cock [in other of words, if the pump is off, and if the water level has fallen and steam is still being produced], he shall immediately throw open the furnace doors and withdraw the fire, under penalty of his being rendered incapable of acting as Engineer or fireman for one year from that time [23].

However, this safe procedure was generally not followed. As a result, **two-thirds** of all steam boiler explosions occurred in the process of getting up steam after leaving a landing—the time when, because the water level had been drawn down during the stop, more of the boiler would have been directly exposed to the fire, become red hot, and then, when the water pump was again actuated, a flood of cold water would come directly into contact with this overheated metal [11, p. 295].

From the vantage point of later developments of the engine and boiler, John Wallace, in his book the *Practical Engineer,* had this to say about this particular problem:

> Taking all this into consideration, it is a wonder that explosions have not been a more frequent occurrence. There was something radically wrong in the former construction of engines. No doubt many good boats have been blown up for want of doctors (*term used later to describe an auxiliary engine that would supply water to the boiler separate from an engine used for propulsion*) to keep up a regular supply of water during long stoppages at wood yards, landing passengers, etc. [20, p. 24].

Problems Operating a Problem-Prone Technology

In reality, the very different problems described thus far would rarely occur in isolation, and frequently they combined to turn an apparently straight-forward system for propulsion into one fraught with dangers, and this complexity was compounded by the lack of training for engineers. Obviously, when the early steam engines were first invented there was no one to learn from, no accumulated body of knowledge in experts who could model safe operations, and no masters from whom apprentices could learn. Everything about steam engines and boilers had to be learned from scratch. Moreover, what had to be learned was *intentionally minimized* by early developers of the high-pressure steam engine such as Oliver Evans.

Endeavoring to make his new steam engines as accessible as possible, Evans first advertised in the May 20th 1812 edition of the *Pittsburgh Gazette* that

> They are less expensive, more simple and durable, occupying infinitely less room, require much less fuel, are thirty times more powerful, and can be conducted by any man of ordinary capacity with less than two months practice [6, p. 187].

However, by April 10th the following year, Evans printed this is *The Commonwealth*:

> A man of ordinary capacity can be taught in two weeks to manage one of these engines completely [6, p. 192].

Twenty-six years later, a report on an explosion written by Professor John Locke spoke directly to this continuing lack of steam engineer training:

> I would ask what has the leisurely enlightened part of the community done for the Engineer? Have you furnished him with a teacher? Have you published for him a Manual to guide him in his important and responsible duties? Have you appointed and paid a qualified person to hear his ingenuous suggestions and to assist him in the perfecting of his inventions? Has our government made the necessary experiments on steam as used in the West, and then put

the result into his hands for his benefit and instruction? None of these things have been . . . [24, pp. 16-17].

The famous steamboat inventor himself, Robert Fulton, wrote an 1851 article describing the type of engineer created in these continuing conditions described by Professor Locke:

> Then there are two classes of practical engineers; one of which is a mechanic and is engaged in building, while the other is employed in running the engine when finished. The first of these may be ignorant of the details of practical operation, and the last is too often ignorant of the first principles of mechanics, either practical or theoretical. This last class do the mischief. They are those who literally kill their thousands, not intentionally, but through ignorance. The history of steam navigation on the Western rivers is a history of wholesale murder and unintentional suicide. We say the general diffusion of knowledge prevents crime—a truth none deny—and we raise money for free schools that all may have the privilege of enjoying them, and at the same time employ engineers who know nothing of the principles of steam or practical mechanics [25, pp. 55-56].

Yet even if the engineers had been given manuals and training, the emphasis on the practical in such training might have not effectively equipped the engineers to deal with problems that were complex and hidden. This process of learning how to operate steamboats was assumed to be like that of previous machines engineers might have encountered at the time—pull the trigger, the lock goes off on the gun; move the foot pedal and the spinning wheel's wheel moves, etc. [26, p. 264]. Knowledge and ability to handle this type of machinery is "experiential" knowledge:

> The essence of expertise is knowing what to do, rapidly and efficiently. The pilot pushes the throttles forward, controlling the nose wheel and rudder to keep the plane on the runway. . . . All this is done with practiced ease and skill, continually integrating numerous sources of information. . . . These are examples of experiential cognition. The patterns of information are perceived and assimilated and the appropriate responses generated without apparent effort or delay. . . . It appears to flow naturally, but years of experience or training may be required to make it possible. . . . Something happens in the world, and the scene is transmitted through our sense organs to the appropriate centers of mental processing. But in the experiential mode, the processing has to be reactive, somewhat analogous to the knee-jerk reflex [27, pp. 22-24].

However, such experiential knowledge and direct mechanical relationships were complicated in the world of steam engines. An intelligence in an operator that could know or understand this world of hidden mechanical relationships was quite different, and is called "reflective" knowledge:

> Reflective thought is very different from experiential thought, even when both are applied by the same people in similar situations. To see this, consider

once again our pilots in their cockpit. . . . Suppose some decision making is required. Suppose the pilots have to plan. Now the situation calls for reflection. Reflective thought requires the ability to store temporary results, to make inferences from stored knowledge, and to follow chains of reasoning backward and forward, sometimes backtracking when a promising line of thought proves to be unfruitful. This process takes time. Deep, substantive reflection therefore requires periods of quiet, of minimal distraction [27, pp. 22-24].

Thus, if the very method of operating the steam boilers and engine most of the time rewarded experiential thinking, emergencies involving an interaction of several aspects of the technology gone awry required reflective thinking. This is exactly what happened with the first steam boiler explosion in the west on the *Washington* in 1816:

The cause of this melancholy catastrophe may be accounted for by the cylinder not having vent through the safety-valve which was firmly stopped by the weight which hung on the lever having been unfortunately slipped to its extreme, without being noticed, and the length of time occupied in wearing before her machinery could be set in motion whereby the force of the steam would have been expended—these two causes united, confined the steam till the strength of the cylinders could no longer contain it, and it gave way with the greatest violence [28, p. 359].

Moreover, the type of control interface—the gauges and stop-cocks—between the operator and the engine and boiler hindered and undercut the engineer's ability to correctly respond when relationships became indirect, hidden, and unclear.

February 24, 1830, Memphis Tennessee, Early Morning

The steamboat *Helen M'Gregor*, with Captain Tyson at her helm and a Mr. Turner as her engineer, was on her way from New Orleans to Louisville, Kentucky. The following was written by a passenger—

The general cry of "a boiler has burst" resounded from one end of the table to the other; and, as if by a simultaneous movement, all started on their feet. Then commenced a general race to the ladies' cabin which lay more toward the stern of the boat. All regard to order or deference to sex, seemed to be lost in the struggle for which should be first and farthest removed from the dreaded boilers. The danger had already passed away! I remained standing by the chair on which I had been previously sitting. Only one person or two staid in the cabin with me. As yet not more than half a minute had elapsed since the explosion; but, in that brief space, how had the scene changed! In that "drop of time" what confusion, distress, and dismay! An instant before, and all were in the quiet repose of security—another, and they were overwhelmed with alarm and consternation. It is but justice to say, that, in this scene of terror, the ladies exhibited a degree of firmness worthy of all praise. No screaming, no fainting; their fears, when uttered, were for their husbands and children, not for themselves.

I advanced from my position to one of the cabin doors for the purpose of inquiring who were injured when, just as I reached it, a man entered at the opposite one, both of his hands covering his face, and exclaiming, "O God, O God! I am lost! I am ruined!" He immediately began to tear off his clothes. When stripped, he presented a most shocking and afflicting spectacle: his face was entirely black; his body without a particle of skin. He had been flayed alive. He gave me his name and place of abode—then sunk in a state of exhaustion and agony on the floor. I assisted in placing him on a mattress taken from one of the berths, and covered him with blankets. He complained of heat and cold as at once oppressing him. He bore his torments with a manly fortitude, yet a convulsive shriek would occasionally burst from him. His wife, his children, were his constant theme: it was hard to die without seeing them; it was hard to go without bidding them one final farewell! Gauze and cotton were applied to his wounds: but he soon became insensible to earthly misery. Before I had done attending to him, the whole floor of the cabin was covered with unfortunate sufferers. Some bore up under the horrors of their situation with a degree of resolution amounting to heroism. Others were wholly overcome by the sense of pain, the suddenness of the fatal disaster, and the near approach of death, which even to them was evident—whose pangs they already felt. Some implored us, as an act of humanity, to complete the work of destruction and free them from present suffering. One entreated the presence of a clergyman to pray for him, declaring that he was not fit to die. I inquired: none could be had. On every side were to be heard groans and mingled exclamations of grief and despair [29, pp. 131-136; 30, pp. 58–69].

Washington City, May 4, 1830— Two and a Half Months Later

In response to the tragedy of the *Helen M'Gregor* (see Figure 13), Congressman Charles Wickliffe of Kentucky rose again to propose steamboat safety legislation. Again it failed to pass, and again the Secretary of the Treasury, this time Daniel Ingham, was instructed to collect information and submit it to a House Select Committee chaired by Wickliffe. The report was given to the Congress two years later, May 18, 1832, in *Report 478*. (At the same time, Ingham pursued a parallel independent investigation done by scientists and mechanics at the Franklin Institute of Philadelphia—the first federally-funded scientific investigation which we will speak of more later.)

Wickliffe's 1832 report was massive—200 pages including reports from around the world, many tables, graphs, and diagrams, reports from reputable scientists (including Walter Johnson who would later direct the ground-breaking strength of materials investigation in the Franklin Institute's *Part II Report* of

Figure 13. The *Helen M'Gregor* explodes.

1836). *Report 478* also included letters and reports from many of those intimately involved with steamboats such as the famous Henry Shreve who had commanded and designed the first river steamboat in the West, the *Washington*, and who, as Superintendent of Western River Improvements, created a boat that removed the infamous snags on the Red River. *Report 478* was also notable for having an introductory section of some nine pages which summarized the findings of the report, offered suggestions on how to correct the problems noted, and contained a ready-to-consider bill. The introduction divided the problems into six parts:

1. Faulty construction of boilers
2. Use of defective material in their construction
3. Long use of boilers which would weaken them
4. Carelessness and want of skill in the engineers
5. Undue pressure in the boiler beyond its capacity
6. Overheated steam caused by insufficient water supply

With *Report 478* in their hands, Congress could no longer complain of too little information and not enough time to consider it, so the obstacle to passage shifted. *Report 478* discussed this new obstacle by including a preamble which described the sense that Congress had at the time of being unable to cope with interstate regulations and the evolution of the technology:

> The distressing calamities which have resulted from the explosion and collapsion [sic] of the boilers of steamboats, the increasing dangers to which the lives and property of so many of our fellow-citizens are daily and hourly exposed from this cause, unite in their demands upon that Government, possessing the competent power and authority, to throw around the lives and fortunes of those thus exposed, all the safeguards which a wise and prudent legislation can give.

> The committee have had more difficulty in determining the extent of the power of Congress to legislate over the subject, than to decide what would be the proper legislation by a sovereign possessing unlimited and unrestricted powers over persons and things. . . .

> Whether the boat or vessel shall be propelled by the wind, or by paddles, or by steam, and, if by steam, whether it be a high or low pressure engine, etc., are questions which it is believed Congress have nothing to do; and if the power were given by the constitution, its exercise might be of doubtful expediency.

> It is better to leave the subject of the application of steam power to the propelling of boats to the sound discretion of those concerned, and to the improvements of the age, than to attempt, by any legislation of ours, to prescribe the particular kind of machinery to be employed.

... It only belongs to legislation to excite, by means of rewards and punishments, that faithful application of those engaged in its use, which will best guard against the dangers incident to negligence [19, pp. 58-59].

According to Wickliffe's report, by May 16th 1832, there had been 52 explosions, 256 killed, and 104 wounded [19, p. 3].

In 1832 the impetus for passage of a bill was again popular pressure arising from accidents as reported in the press. Moreover, like the earlier Vinton bill, Wickliffe's bill looked at both the technological problems as well as the problems caused by faulty and negligent decisions by operators. To deal with the technological problems, Wickliffe's bill offered the same solutions as did Vinton: periodic inspections of boilers and licenses with the enforcement of fines, but Wickliffe's bill added hydrostatic pressure testing of boilers. To deal with the faulty decisions by engineers and captains, Wickliffe's bill ignored Vinton's emphasis on the safety valves and tampering, and instead focused on the operational problem of stopping and starting the engine and boilers during the stops on a steamboat passage; the bill would require that the water pump be kept pumping and the safety valve weight be diminished so as to decrease steam pressure in the boilers. In addition, in a bill directed to boiler explosions, Wickliffe also included four elements to insure general boat safety: long boats for passengers, fire hoses and equipment, bow running lights, and a requirement that when boats would pass in the night, the boat descending would shut off steam until clear.

This time the counter pressure against passage of the bill moved from not having enough information to complaints about Congress not having the power to enact interstate commerce regulation. In addition, the belief in the continuing progress of technical evolution moved Congress to be wary of impinging upon the "improvements of the age." Finally, the inclusion of general safety aspects tended to confuse the focus upon steam boilers and their control. The bill, this time for wholly new reasons, was not passed.

Endnotes

1. "A Song Dedicated to the Survivors of the Wreck of the *Golden Gate*," Words by Frank Soule, Music by P. R. Nicholls, Charles F. Robbines & Co, 1862, Lester S. Levy Collection of Sheet Music, Special Collections, Milton S. Eisenhower, Johns Hopkins University, Box 179, Item 75.
2. R. John Brockmann, *From Millwrights to Shipwrights to the Twenty-first Century: Explorations in a History of Technical Communication in the United State*s, Hampton Press, Norwood, New Jersey, 1998.
3. Louis C. Hunter, *A History of Industrial Power in the United States, 1780–1930, Volume Two: Steam Power*, University Press of Virginia Published for the Eleutherian Mills–Hagley Foundation, Charlottesville, Virginia, 1985.
4. Harlan I. Halsey, The Choice Between High-Pressure and Low-Pressure Steam Power in America in the Early Nineteenth Century, *The Journal of Economic History, 41*:4, pp. 723-744, December 1981.

5. Charles Joseph Latrobe, The Western Steamboat, in *Before Mark Twain: A Sampler of Old, Old Times on the Mississippi*, John McDermott, editor, Southern Illinois University Press, Carbondale, Illinois, 1968.

6. Greville and Dorothy Bathe, *Oliver Evans: A Chronicle of Early American Engineering*, Arno Press, New York, 1972.

7. Coleman Sellers, Oliver Evans and His Inventions, *Journal of the Franklin Institute, 122*:1, July 1886. Some of this opposition was drummed up by Evans competitors such as Col. John Stevens who wrote articles in the journals of the times claiming that high pressure steam was too dangerous to be used.

8. Dorothy Gregg, *The Exploitation of the Steamboat: The Case of Colonel John Stevens*, Dissertation, Ph.D. in Political Science, Columbia University, New York, 1951.

9. Charles Dickens, *American Notes*, Wilson & Co., New York, 1842.

10. James Renwick, *Treatise on the Steam Engine*, Carvill & Co., New York, 1839.

11. Louis C. Hunter, *Steamboats on the Western Rivers: An Economic and Technological History*, Harvard University Press, Cambridge, 1949.

12. George W. Pierson, *Tocqueville and Beaumont in America*, Oxford University Press, New York, 1938.

13. E. W. Gould, *Fifty Years on the Mississippi*, Nixon-Jones Printing Co., St. Louis, Missouri, 1889. "A curious fact was ascertained by a committee of gentlemen, who were appointed a few years ago, by a number of steamboat owners to investigate the whole subject. They satisfied themselves, that although the benefits conferred on our country, by steam navigation, were incalculable, the stock invested in the boats was, as a general rule, a losing investment. In few cases, owning to fortuitous events, or to the exercise of more than usual prudence, money has been made; but the instances are so few as not to effect the rule. One gentleman, who has been engaged for years in the ownership of steamboats and has been particularly fortunate, in not meeting with any loss by accident, assured the writer, that his aggregate gain, during the whole series of years, was only about six per cent per year, on the capital invested."

14. Steamboats on the Western Waters, *Journal of the Franklin Institute, 14*:5, November 1834.

15. Eric F. Haites and James Mak, Social Savings Due to Western River Steamboats, *Research in Economic History, 3*, pp. 263-304, 1978. Haites and Mak note that although the return on investment in steamboats was low, they far exceeded those in canals and railroads.

16. William King, *Lessons and Practical Notes on Steam, the Steam-Engine, Propellers, etc., etc*, 4th edition, New York, 1863.

17. *St. Louis Times-Journal* qt. in *Cincinnati Gazette*, September 30, 1879.

18. Charles Edward Russell, *A-Rafting on the Mississip'*, The Century Co., New York, 1928.

19. Report No. 478, Steamboats, May 18, 1832, in *Reports of Committees of the House of Representatives at the First Session of the Twenty-Second Congress, Begun and Held at the City of Washington, December 7, 1831*, Duff Green, Washington, 1831.

20. John Wallace, *Practical Engineer: Showing the Best and Most Economical Mode for Modeling, Constructing and Working Steam Engines, Written in a Plan, Concise and Practical Style, and Designed Especially for Practical Engineers, Steam Boat Captains, and Pilots*, Kennedy & Brother, Pittsburgh, Pennsylvania, 1853.

21. Right from the beginning when Oliver Evans had first designed the high-pressure engine, he had equipped it with two safety valves. By 1838, this was no longer standard.

22. Written and air adapted by Charles Robertson, London: Augener & Co, Lester S. Levy Collection of Sheet Music, Special Collections, Milton S. Eisenhower, Johns Hopkins University, Box 181, Item 79.

23. "Document in relation to a mode of preventing the explosions of steam boilers. 1837, Dec. 29 referred to the Select Committee to whom was referred the Bill S. 1. 25th Congress, 2nd Session, p. 5. National Archives, Center for Legislative Archives, SEN25A-B1—B6, D19 Boxes 1, 4, 6, 8, and 27.

24. *Report of the Committee Appointed by the Citizens of Cincinnati, April 26, 1838, to Inquire into the Causes of the Explosion of the Moselle, and to Suggest such Preventive Measures as May Best Be Calculated to Guard Hereafter Against Such Occurrences*, Alexander Flash, Cincinnati, Ohio, 1838.

25. Robert Fulton, What Constitutes an Engineer? *Journal of the Franklin Institute, 21*, pp. 55–56, 1851.

26. James Paradis, Text and Action: The Operator's Manual in Context and in Court, in *Textual Dynamics of the Professions*, Charles Bazerman and James Paradis, Editors, University of Wisconsin Press, Madison, Wisconsin, 1991. Paradis suggests that users sometimes learn to operate technologies based on a stock of generic images. "User strategies can be built upon the stock of generic images that, as Boulding has argued, is shared by society. Everyone 'operates' a screwdriver or flashlight without having to be instructed. In a thriving consumer society, this kind of intuitive operational know-how based upon socially shared imagery must be widely available. Many lawnmower purchasers can by mere inspection decode the fraction of the technology necessary to operate the instrument satisfactorily. Hence, the sport of dispensing with the manual: When all else fails, the saying goes, 'consult the manual.' The highly accessible technologies common in a consumer society are thus based on a social substrate of shared generic imagery, a kind of Platonic world of idealized forms and processes that is presumably the product of elementary and second school education, supplemented by television culture." However, in 1830 rural America, before the consumer society and its "shared generic imagery," was there any "shared generic imagery" appropriate to this new steam engine technology?

27. Donald A. Norman, *Things That Make Us Smart: Defending Human Attributes in the Age of the Machine*, Addison-Wesley Publishing Co., Reading, Massachusetts, 1993.

28. Horrid Accident, *Weekly Recorder*, Chillicothe, Ohio, June 13, 1816.

29. S. A. Howland, *Steamboat Disasters and Railroad Accidents in the United States*, Dorr, Howland & Co., Worcester.

30. Joseph L. Cowell, Joe Cowell Recalls A Voyage in the Helen M'Gregor in 1829; or, Twelve Days Confinement in a High Pressure Prison, in *Before Mark Twain: A Sampler of Old, Old Times on the Mississippi*, John McDermott, Editor, Southern Illinois University Press, Carbondale, Illinois, 1968.

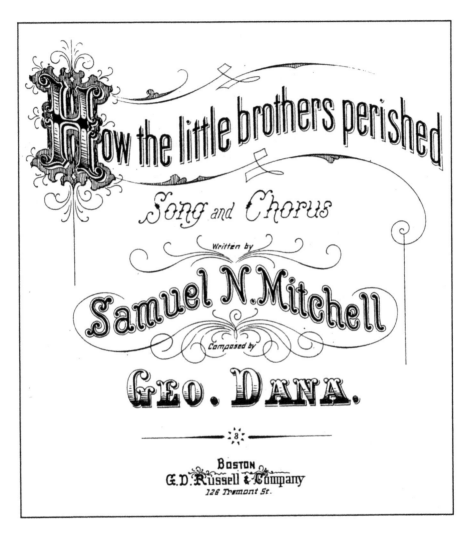

Figure 14. "How the Little Brothers Perished" [1; 2, p. 61].

In the morning dark and dreary,
Of that fatal August day,
Cast among the floating timbers,
Hard they struggled with the spray,
Not a whisper, not a murmur,
Came from either little tongue,
As they left the sinking vessel,
Let it thus to die so young.
Like brothers brave and true they perish'd,
Clasped in one another's arms,
While around the air was ringing,
Ringing with the mad alarms.

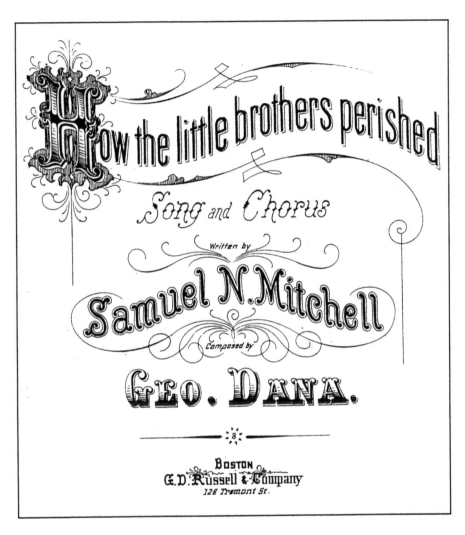

Figure 14. "How the Little Brothers Perished" [1; 2, p. 61].

In the morning dark and dreary,
Of that fatal August day,
Cast among the floating timbers,
Hard they struggled with the spray,
Not a whisper, not a murmur,
Came from either little tongue,
As they left the sinking vessel,
Let it thus to die so young.
Like brothers brave and true they perish'd,
Clasped in one another's arms,
While around the air was ringing,
Ringing with the mad alarms.

CHAPTER 3

Steamboats, The Presidency, and Public Opinion

Red River, May 19, 1833, Early on a Spring Sunday Morning

On the night of the 19th, the Hon. Josiah Johnson, senator from Louisiana (see Figure 15), was killed along with 13 others, and the Hon. Edward White, also of Louisiana, was injured when the steamboat *Lioness* blew up (see Figure 16) [3, pp. 83-85]. The *Baltimore American* once more called for action:

> The number of fatal accidents on the Mississippi—particularly the disastrous one by which Senator Johnson lost his life—imperiously call for some legislative interference [4, p. 153].

December 3, 1833—President Jackson's State of the Union Message to Congress

As a Westerner from Tennessee, Andrew Jackson personally knew the paradoxical relationship Westerners had with the steamboats—they needed the steamboats with their power and carrying capacity, but they feared their unreliability. Thus, in a letter to his son about operations on their own jointly owned Tennessee farm, Jackson wrote:

> There have been so many Steam Boat accidents of late, that I would not ship the cotton on board of any boat who had not an experienced engineer, and a careful and experienced Captain. Ship with none that will have any combustible on their upper deck. I advise you to be careful in selecting a good Boat with careful and experienced engineer and commander . . . [5, p. 227].

Now that a senator from Louisiana was dead and one injured in a steamboat catastrophe, Jackson offered a solution in the first State of the Union message of his second term:

> The many distressing accidents which have of late occurred in that portion of our navigation carried on by the use of steam power, deserve the immediate

43

Figure 15. Senator Johnson.

and unremitting attention of the constituted authorities of the country. The fact that the number of those fatal disasters is constantly increasing, notwithstanding the great improvements which are everywhere made in the machinery employed, and in the rapid advances which have been made in that branch of science, show very clearly that they are in a great degree the result of criminal negligence on the part of those by whom the vessels are navigated, and to whose care and attention the lives and property of our citizens are so extensively entrusted.

That these evils may be greatly lessened, if not substantially removed, by means of precautionary and penal legislation, seems to be highly probable: so far, therefore, as the subject can be regarded as within the constitutional purview of Congress, I earnestly recommend it to your prompt and serious consideration [6].

In Jackson's speech, the complexity of the problem described earlier by Vinton and in Wickliffe's reports and bills—that the solution needed to address both the technological and operational problems—was simplified and made a case

Figure 16. The *Lioness* explodes.

of criminal negligence. The optimism offered by both Crawford in his 1825 report and in Wickliffe's report that the natural evolution of the technology would solve the technical problems was now replaced by technological pessimism— "notwithstanding the great improvements which are everywhere made in the machinery employed, and in the rapid advances which have been made in that branch of science. . . ."

By directly addressing the steamboat problem himself, President Jackson qualitatively altered the discussion initiated by Vinton and carried on by Wickliffe. No longer was the problem the focus of little known members of the House of Representatives, but now it was the focus of national attention by a President and, soon, one of the most powerful senators of the day, Daniel Webster.

Usually Jackson, the Democrat President from the west, and Daniel Webster (Figure 17(a)), the Whig Senator from Massachusetts, did not see eye-to-eye and had numerous public arguments. Yet, in early 1833, Webster joined Jackson in common cause for preserving the union against John C. Calhoun (Figure 17(b)) and his nullification efforts to undermine the union of the United States [7, p. 530; 8, Chapter 22]. Relations between Jackson and Webster during this policy fight improved so much that there were rumors Webster might become Jackson's chosen successor.

On December 14th Jackson's long-time friend and fellow Tennessee-an, Senator Felix Grundy, brought word to Jackson from the Senate that "an

(a)

(b)

Figure 17. Senators Daniel Webster (a) and John C. Calhoun (b).

arrangement could be made with Mr. Webster and his friends to win control of crucial committees for the administration" [9, p. 17]. It was an offer that had Webster's presidential ambitions all over it; moreover, a Webster biographer noted that this was the very moment in which Webster came closest to making it to the White House [10, p. 142]. Only the vehement protestations of Jackson's friend and Vice President, Martin Van Buren, stopped the "arrangement" so that Van Buren himself could assume Jackson's mantle in the next election. In his autobiography, Van Buren triumphantly noted: "Between neither of these gentleman and myself was the subject ever reviewed" [11, p. 679].

Yet, even if the entire "arrangement" could not be consummated, it was possible that Webster reasoned he could demonstrate to Jackson what he, Webster, could do with one paragraph out of Jackson's 13-page State of the Union message. Thus, on December 23, 1833, Webster rose to propose the following piece of legislation based upon Jackson's single paragraph initiative:

> **Resolved,** That the Committee on Naval Affairs be instructed to inquire into the expediency of passing a law for preventing, as far as may be, accidents to vessels employed in the foreign or coastwise commerce of the United States from the explosion of steam [12, p. 765].

However, Webster did not just leave the issue there for committee considerations, but proceeded to outline a bill:

> The history of the country for the last three years presented a most startling list of accidents from explosions by steam, and the general opinion seemed to be

that they arose, in many instances, from very culpable negligence, but in some from a more positive criminal offense than negligence, to wit: steamboat racing. This was a most unpardonable offense as tending immediately to the destruction of life, by an agent so salutary and useful when under proper restraint, but awfully destructive and calamitous when in inexperienced hands.

Some general ideas upon the proper remedy, however, have occurred to me which I will suggest for consideration.

The law, as it seems to me, might be of twofold character. I might prescribe certain regulations, the violation of which, whether accidents happen in consequence or not, should incur a penalty; and it might further provide, that, in the case of accident, although all prescribed regulations should have been previously complied with, yet, if the accident happened from culpable negligence at the moment, that negligence should be severely punished. As to previous and prescribed regulations, the first and most important, doubtless, should be that every boiler intended for a steamboat should be tried and proved by some public authority, and restrained, in its future use, to one-third, or at most one-half, the degree of pressure or tension which it should have been proved to be capable of bearing. A safety valve, out of the reach of the fireman, a proper apparatus to show at all times the level of the water, and also the intensity or height of the steam, and this apparatus so arranged that its indications may be seen from without, are among the preventive remedies to which the attention of the committee will probably be called.

But I look with more confidence of beneficial results from certain other provisions, which I trust will receive the consideration of the committee. Fully believing that these accidents generally result from negligence, at the time, by those who have the charge of the engine, penalties, I think, ought to be enacted against such negligence, and legal means provided, by which when lives are lost by such occurrences, an immediate inquisition, investigation, and trial should be secured, and the culpable negligence, if there be such, adequately published. And in the first place, I think the boat itself should be made subject to forfeiture, whenever lives were lost through the negligence of those conducting it. There is nothing unreasonable in this; analogous provisions exist in other cases. The master of a merchant ship, for instance, may forfeit the ship by a violation of the law, however innocent the owners may be, even though that law be only a common regulation of trade and customs. There is at least quite as much reason for saying that whoever builds or buys a steamboat, and proposes to carry passengers therein for hire, shall be answerable to the amount of the value of the boat, for the sobriety, diligence, and attention of those whom he appoints as his agents to navigate it, as there is, in revenue cases, to impose such liability for smuggling, or illegal landing of goods. To enforce this liability, I should propose, that whenever an explosion takes place, causing the loss of the lives of passengers, the boat should be immediately seized by the collector of the district, and the persons navigating her detained for examination; a trial should be had, and unless it should appear on such trial that all legal requirements had been previously complied with and were observed at the time, and further, that the accident was one which no degree of attention could have foreseen or prevented, the boat should be forfeited and the persons having charge at the time should be punished. It is no unreasonable hardship, in such cases, to throw the burden of proof on those who are entrusted with the navigation and management of the boat. They should be able to make out a clear case of actual attention, skill, and vigilance, or else forfeiture ought to follow. It is a very high trust to have charge of that which is so potent to destroy life, and which, when negligently treated, is so likely to destroy it. Of course, all unnecessary delay, expense, or trouble should be avoided. The property seized might be restored, on bonds, as in other cases of seizure, pending preparation and trial; and every indulgence allowed, in the forms and modes of proceedings, compatible with the great end of an immediate investigation and a prompt decision.

It is evident that, for many reasons, a judicial investigation will seldom be had, in these cases, unless it be instituted by public authority; and I do not think any provisions will be adequate which do not secure such investigation, whenever the loss of life happens [12, pp. 765-766; 13, p. 647].

Following Jackson's emphasis on criminal negligence—

Fully believing that these accidents generally result from negligence, at the time, by those who have the charge on the part of those by whom the vessels are navigated, and to whose care and attention the lives and property of our citizens are so extensively entrusted [12, pp. 765-766; 13, p. 647].

—the penalties of Vinton and Wickliffe's simple fines grew in Webster's bill to include mandatory trials and boat confiscation. This civil liability approach was the means by which Webster foresaw most improvement could be had.

However, Webster also returned to the technological side of Vinton and Wickliffe's solutions by again bringing up the idea of testing and licensing boilers to a particular pressure, and to require two separate safety valves, one of which would be out of the reach of engineer tampering. Webster also added a new element to the technological solutions—the need for more indicators to tell engineers the level of the water and of the steam. Further, he called for such instrumentation to "be read from without" intending that if any rash actions of an engineer raised the pressure levels too high, it could be easily observed, and that the self-interests of other crew members, captains, owners, and passengers would act to correct the engineer's actions. Hence, Webster knew it wasn't all negligence that caused the explosions, and Webster attempted to create provisions in a number of areas to make the steam technology itself more manageable.

By weighing into the debate, Webster himself contributed to the new political pressures and counter-pressures on the steamboat regulation bill. Pressure for the bill still came from popular pressure stoked by the vivid accident accounts in the newspapers. But in addition, Webster brought to bear the sensibilities of the rising power of the Whig Party, a party that fervently believed that "Popular government follows the track of the steam-engine and the telegraph" [13, p. 647]. Webster himself had been long involved in steamboats and steamboat litigation.

Webster loved to crow that in the 1824 Supreme Court Case, *Gibbons v. Ogden,* which ended the monopoly of Fulton's steamboats on the Hudson and opened up steamboat competition throughout the United States:

> The opinion of the court was little else than a recital of my argument. The chief justice told me that he had little to do but to repeat that argument as that covered the whole ground [14, p. 285; 15, p. 27].

Later, in 1834 when a Whig member of the Virginia House of Delegates wanted to write to someone to urge steamboat safety, he wrote to Webster—"I hope you may get something done in regard to the safety of running steam boats" [16]. And, in 1840, in a 46-page *Letter to the Hon. Daniel Webster on the Causes of the Destruction of the Steamer Lexington,* Webster's central role in steamboat safety was again noted:

> The particular interest which you, Sir, have manifested in the safety of travelers by your efforts to legislate in their behalf, and for which I feel that I may, in the name of thousands, offer you the most sincere thanks, induces me to take the liberty of addressing my remarks to you [17, p. 6].

Intimately involved with steamboats and steamboat technology, motivated by the expedience of the moment in common cause with Jackson, and sincerely believing in the Democratic effects of technology, Webster wrote:

> It is an extraordinary era in which we live. It is altogether new. The world has seen nothing like it before. I will not pretend, to discern the end; but for scientific research into the heavens, the earth, and what is beneath the earth; and perhaps more remarkable still for the application of this scientific research to the pursuits of life [18, p. 15].

Senator Benton, the Democrat Senator from Missouri, rose to speak on the Webster/Jackson steamboat initiative as a member of the opposition party. Benton said [19, p. 4] that although he did not oppose the resolution, he urged that the consideration of such steamboat explosions be taken up by the Judiciary Committee rather than the Naval Affairs Committee [20] since the private waters of the states were involved, and it might be a case of interference with their sovereignty. Moreover, Benton said:

> notwithstanding so many calamitous accidents had occurred, his acquaintance with the captains and owners of steamboats were, with few exceptions, men of the highest integrity. Further, he had never met with any accident on a steamboat despite the fact that he traveled widely; upon boarding he was always careful to inquire whether the machinery was in good order [21, p. 12].

Webster succeeded in this skirmish, and the bill's consideration was moved to the Naval Affairs Committee. And, where he knew that even if he did not chair the Committee, his close Whig ally, Samuel Southard of New Jersey, did. Thus, in a small demonstration of his skills, Webster illustrated what "arrangements" he could have wrought for Jackson.

A bill was brought to the Senate floor from the Committee six months later on June 17, 1834. Many of the elements of Vinton's and Wickliffe's earlier bills were included in the Naval Affairs Committee's draft bill. For example, the Naval Affairs Committee Bill, like the Vinton and Wickliffe bills, proposed periodic testing, keeping the water pump operating when the vessel was stopped, shutting off steam when passing ascending ships, having fire-fighting equipment, etc. However, there were some unique inclusions and exclusions in the Naval Affairs Committee bill. The Naval Affairs bill had two items on gunpowder transportation aboard steam vessels that were probably reactions to the finding that the *Lioness* explosion in May 1833—the one that killed a senator and motivated President Jackson to propose a law in his State of the Union message—was due to accidental detonation of gunpowder barrels by fire from the steam engine. Neither these gunpowder safety items, nor the testing and licensing of individual steam engineers ever appeared in any other proposed bills [22, p. 153]. The Naval Affairs Committee bill, however, did not mention safety valves at all even when Webster himself called for them when he introduced the legislation.

By June, when the Naval Affairs Committee reported out their bill, the honeymoon between Jackson and Webster was over. In the interim, Webster replaced the soft words of "arrangement" with ones in which he lambasted Jackson as "a military dictator . . . in which Rome had no better models" [23, p. 247].

The bill was tabled.

Neither Jackson nor Webster had really suffered any political fallout with the failure of this steamboat initiative. It was, after all, but one paragraph in Jackson's 13-page State of the Union, and, for Webster, he had not really proposed a specific bill but had only outlined one, and then had thrown it to Southard and his lesser known colleagues in committee. Moreover, for his opposition to Webster's initiative, Thomas Benton of Missouri did not suffer either; he was nominated to be Van Buren's running mate by a Democrat convention later that same Fall [24, p. 203].

But What About the Public Pressure for Steamboat Safety?

In these first three Congressional ventures into steamboat regulations there was a fairly direct relationship between explosions, their lurid coverage in newspapers, and attempts to pass bills in Congress. Vinton, in 1824, spoke of it:

> . . . a melancholy instance had recently occurred near New York; the sensation produced by the news of which in this house had not yet subsided [25, p. 202].

And the *Congressional Globe* of Washington City noted it about Webster in 1833:

> He (Webster) had brought forth the resolution [his 1833 attempt] because
> so much terror existed in the public mind as greatly to impede that most
> comfortable and useful mode of traveling [26, p. 1].

Such subjective impressions of the relationships between explosions and
legislative attempts can be empirically demonstrated by plotting a graph of deaths
and injuries aboard steamboats from 1824 to 1833 and noting that legislative
attempts usually occurred at the peaks when deaths, injuries, and explosions were
most often reported in the papers. In Figure 18, one can see the fairly direct
relationship between the reports of these "melancholy instances" and the move
on the floor of either the House (in 1824 and in 1830) or in the Senate (in 1833)
to pass some protective legislation.

Even when the death and injuries are adjusted for the increased number of
steamboats and passenger miles between 1816 and 1838 [27, p. 183], the
relationship between an increase in injuries and deaths and the various legislative
efforts still operates as seen in Figure 19 [28, p. 80].

Thus the public response to the explosions was a primary driving force
behind the legislation. Yet public response was based less on objective facts than
on passions, for if it were raw numbers of dead and killed were tallied, the public
and Congress should have been looking at safety aboard sailing ships, stages, and
carriages. This passionate element of what was driving the public opinion was
noted a number of times by those who looked a bit more dispassionately at the
data, but it never quite affected the public . . . or the Congress.

Explaining his observations on a table he constructed which laid out the
explosions of steam boilers in 1830, W. C. Redfield of the Steam Navigation
Company wrote to the Secretary of the Treasury as part of *Report 478*:

> Although this is a melancholy detail of casualties, yet it seems less formidable
> when placed in comparison with the ordinary causes of mortality, and espe-
> cially when contrasted with the insatiate demands of intemperance and ambi-
> tion. It is believed that it will appear small when compared with the whole
> amount of injury and loss which has been sustained by traveling in stages and
> other kinds of carriages. More lives have probably been lost from sloops
> and packets on the waters of this State [New York] since the introduction of
> steamboats, than by all the accidents in the latter, though the number of
> passengers has been much smaller. In one case that occurred within a few
> years, thirty-six persons were drowned on board a sloop in the Hudson river,
> and many instances, involving the loss of a smaller number of lives; and
> one case occurred not long since, on Long Island sound, which resulted in the
> loss of twelve or fourteen individuals [29, p. 21].

In April of 1832, Captain Henry M. Shreve, whom many regarded as the hero
of western steamboat transportation [30, p. 85], concurred with Redfield in the
following words:

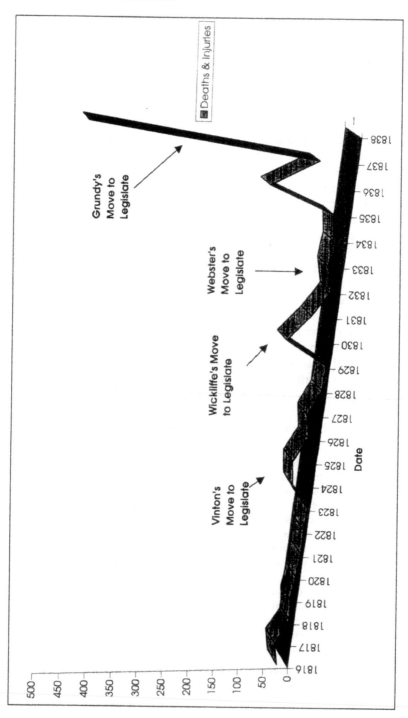

Figure 18. Deaths and injuries from boiler explosions (1816-1838) with legislative initiatives noted.

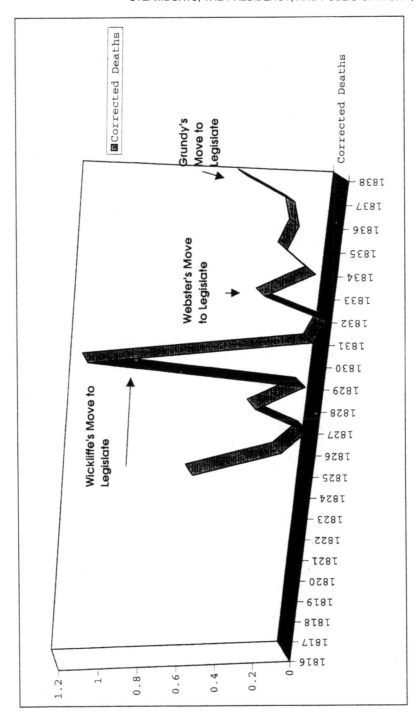

Figure 19. Deaths and injuries from boiler explosions (1824-1838) with legislative initiatives noted using Denault's corrections.

Mr Redfield's remarks, in the tenth page of his communication, are doubtless correct. If Congress would inquire into the loss of human life by drowning, &c., from vessels navigated within her territory, since the twentieth of June, 1816, the date of the first explosion, to this time, it will probably far exceed the number lost by explosions of steam boilers, which has been, in that period, two hundred and sixty-three, caused by fifty-two explosions, and the comparative number of passengers on steamboats, since that date, must exceed those on all other crafts by at least three to one [29, p. 59].

Redfield's and Shreve's comment proved persuasive to the Wickliffe Committee because in the Committee's summary to *Report 478* the Committee wrote:

The melancholy incidents which have occurred by the explosion of many of the boilers of steamboats in the waters of the United States; the shock which is universally felt on these occasions, had impressed the committee with an opinion that the destruction of human life had been much greater than it turns out to be upon further and more minute investigation [29, p. 3].

Yet the Wickliffe Committee returned to the sway of public passions in its ultimate paragraph:

The committee, therefore, report a bill, and earnestly entreat the House to give to it that consideration which the nature and importance of the subject demand; and, if the same shall be approved, that the House will pass upon it at the present session, and, as far as possible, give quiet and repose to the public mind, which has been so long and so anxiously directed to this subject [29, p. 6].

Sixteen years later, in 1848, the Commissioner of Patents, Edmund Burke, again concurred with Redfield's and Shreve's observations concerning the lack of objectivity in public opinion regarding the explosions:

There is something in the appalling nature of steam boiler explosions which strikes public attention, and has given rise to an impression that steamboats and railroads are more dangerous modes of conveyance than any others. It is to be regretted that no direct means of comparison though a series of years, between the losses by ordinary navigation, and those by steam navigation, are in the possession of this Office. . . . But a general and somewhat loose comparison can be made, which may serve to correct a false impression which undeniably exists with regard to the comparative safety of the two-modes of travel. It appears from a statement contained in a memorial addressed to Congress on this subject, that in the year 1839 the number of American vessels lost by ordinary navigation was 1059, and 179 lives were destroyed. Thus the number of lives lost in that month is nearly double the average annual loss of life by steamboat explosions, as deduced from the foregoing calculations. . . . Yet so much terror has been excited in the public mind, by accidents of this kind, that the prevention of them has (and no doubt properly) been considered by those nations which have made most use of the powerful and useful, though dangerous agency of steam, as calling for special legislative interference [31, p. 3].

Finally, Louis Hunter, in his 1949 landmark study *Steamboats on the Western Rivers,* also concurred when he noted:

> Public sentiment on the subject of steamboat accidents, needless to say, was not based primarily on statistical and comparative studies. The cumulative figures of steamboat losses were not generally available, and the data on the comparative risk of stagecoach, sloop, and steamboat, even if accurately determined and widely published, would not have materially affected the issue. What aroused public opinion and moved legislative bodies was less the cold calculation of total losses and relative risks than the shock of individual disasters which did not occur at an exotic distance, but frequently at one's very doorstep. Many were inclined to accept the mid-century observer who declared: "The history of steam navigation on the Western Rivers is a history of wholesale murder and unintentional suicide" [32, p. 522].

Some of the reasons for the heightened awareness in antebellum society of the steamboat explosions were because most of those being injured were innocent parties, the passengers. It is interesting to note that the mortality among the crew was not a focus of Vinton's quick report in 1824 (*Report 125*) which had in the title—"a bill for regulating of Steam Boats, and for the security of *passengers* therein." Wickliffe's (1830) law was similarly focused on the innocent parties, "to provide for the better security of *Passengers*."

The heightened awareness could also have arisen because the boiler-explosions killed indiscriminately, quickly, and in large numbers—at the time, there weren't any other cases of such large losses of life occurring as a result of human endeavors. Hurricanes and floods may have killed more, but, at this time in the early 19th century, nothing else created by humans could wreak such havoc outside of war.

Perhaps the most important reason, however, for the heightened awareness about steamboat explosions was the unresolved ambivalence in the American imagination that cutting-edge technology such as steam engines was producing. Here in an era before the mass-production of any other technology, before sewing machines or McCormick reapers had been successfully marketed, here was magic from the minds of man that as one preacher put it:

> . . . It is to bring mankind into a common brotherhood; annihilate space and time in the intercourse of human life; increase the social relations . . . multiply common benefits . . . and by a power of unknown kindness, give to reason and religion an empire which they have but nominally possessed in the conduct of mankind [33; 34, p. 212].

or, as another put it:

> Such is the general picture; examine it more carefully for a moment. See how true children of the sun, heat, light and electricity have been delivered into the

hands of man, as bondservants, obedient to his call . . . With consummate skill
the marriage of water and heat was effected. The child of that marriage has
grown to be a Herculean aid to onward-moving humanity. Certainly steam is a
benefactor to the race [35; 36, p. 76].

The poet Walt Whitman was in a similar frame of mind when he wrote after
visiting the engine room of a ferryboat:

It is an almost sublime sight that one beholds there; for indeed there are few
more magnificent pieces of handwork than a powerful steam-engine at work!
[37, p. 811].

Oliver Wendell Holmes wrote a poem in 1837 entitled "The Steamboat," and the
third stanza was as follows:

Now, like a wild nymph, far apart
She veils her shadowy form,
The beating of her restless heart
Still sounding through the storm;
Now answers, like a courtly dame,
The reddening surges o'er,
With flying scarf of spangled flame,
The Pharos of the shore [38, p. 216].

The depth to which steamboats reached into the antebellum imagination could
be seen no better than by an observation that the visiting Swedish novelist
Fredrika Bremer made when she observed a period of undirected drawing at
a boys' school. She discovered most of the children sketched on their slates
"smoking steam-engines or steam-boats, all in movement" [39, p. 41].

Yet, at the very same moment when this delight, awe, and magic of steamboats filled the American imagination, dozens of lives were being lost. The
experience of such disastrous, unintended effects was new and shocking. The
newspaper stories of the havoc and mayhem of these steamboat explosions was so
popular with the public that they were:

• collected into a sub-genre of disaster literature such as *The Mariner's
  Chronicle: Containing Narratives of the Most Remarkable Disasters at Sea*
  (1834) [40], Charles Ellms's *Shipwrecks and Disasters at Sea* (1836) [41],
  S. A. Howland's *Steamboat Disasters and Railroad Accidents in the United
  States* (1846) [42], and *Lloyd's Steamboat Directory and Disasters on the
  Western Waters* (1856) [43];
• put into sheet music so that the melancholy could be shared in the parlor with
  such songs as those depicted on each new section of this book, but also
  including "The Bell of the Atlantic" [44, 45]; "Little Commodore; or The
  Rescued Boy of the Atlantic" [46]; "Lost on the Lady Elgin" [47]; "Grace
  Darling" [48]; "The Lost Steamer" [49]; "The Gold Dust Fire" [50, p. 41]; and

• depicted on the covers of children's jigsaw puzzles of steamboats which did not advertise "Steamship Jigsaw Puzzles" but rather "Blown Up Steamboat" (see Figure 20).

The explosions even became a point of comparative advertising; the Steam Navigation Company Line advertised the fact that they had towed "safety barges" for women—a barge towed by lines behind the steamboat (see Figure 21) and on which there would be no engine and thus no possibility of explosion [51, pp. 58-65; 52, pp. 10, 90-91]. Newspapers of the time included advertisements with pictures of the barges and rhetorical questions such as "Why sleep over a volcano?" [51, p. 59] One advertisement had the following appeal:

> Passengers on board the Safety Barges will not be in the least exposed to any accident by reason of the fire or steam on board the steamboats. The noise of the machinery, the trembling of the boat, and the heat from the furnace, boilers, and kitchen, and everything which may be considered unpleasant on board a steamboat are entirely avoided [53, p. 38].

Figure 20. Cover of "Blown Up Steamboat" jigsaw puzzle.

Figure 21. Safety barge in tow on the Hudson [53, p. 39].

Just as the fascination with the positive aspects of the engines entered deep into the American imagination, so did the dangers of these engines. For example, these dangers gave rise to tall tales such as:

Talk about Northern [northern Mississippi river boats] steamers, it don't need no spunk to navigate them waters. You haint bust a boiler for five years. But I tell you, stranger, it takes a man to ride one of these half alligator boats, head on a snag, high pressure, valve soddered down, 600 souls on board and in danger of going to the devil [54, p. 100].

Or this one collected by W. H. Milburn:

The Coolest Feller

Jim Smith was fur sartin the coolest feller in the country. He had a little one-room cabin on the banks of the Mississip', an' one night as he sot cleanin' his rifle, one side o' the fire, an' Matty, his wife on the other, knittin', the' cum a terrible explosion on the river, an' the next minute something' cum plum through the roof an' dropped at their feet, right between 'em, without disturbin' either.

Jim went on a-cleanin' his gun an' Matty, she kep' knittin'.

The stranger—fur it was a man—was a little dazed at fust, but gittin' up, he squinted at the hole in the roof an' says he, "Well, my man, what's the damage?"

Jim put down his rifle, took a careful look at the hole, figured a while an', says he, "Ten dollars."

"You be hanged," said the traveler.

"Last week I was blown up in another steamboat, opposite St. Louis, and fell through three floors of a new house, and they only charged me five dollars.

No, no, my man, I know what the usual figure is in such cases. Here's two dollars; if they won't do, sue me as quick as you please!" [55]

In addition to such folk tall-tales, the dangers of steamboat explosions also popped-up in Nathaniel Hawthorne's literary, short story "The Celestial Railroad" (1843). In that story, after supposedly conveying his pilgrims to heaven aboard a steam railroad, Mr. Smooth-it-away delivered his charges to the jaws of hell in the last few paragraphs by:

> A steam ferry boat, the last improvement on this important route, lay at the river side, puffing, snorting, and emitting all those other disagreeable utterances which betoken the departure to be immediate [56, p. 56].

A national ambivalence between an attraction to steamboat power and a repulsion to its dangers only drew "the public mind, which has been so long and so anxiously directed to this subject" [57, p. 6]. Michael Chevalier wrote about the America of the 1830s in his book *Society, Manners and Politics in the United States: Being A Series of Letters on North America*. In its pages, Chevalier described how Americans dealt with this deep ambivalence about steamboats:

> In the West, the flood of emigrants, descending from the Alleghanies, rolls swelling and eddying over the plains, sweeping before it the Indian, the buffalo, and the bear. At its approach the gigantic forests bow themselves before it, as the dry grass of the prairies disappears before the flames. It is for civilization, what the hosts of Ghengis Khan and Attila were for barbarism; it is an invading army, and its law is the law of armies. The mass is everything, the individual nothing.
>
> Woe to him who trips and falls! He is trampled down and crushed under foot. Woe to him who finds himself on the edge of a precipice! The impatient crowd, eager to push forward, throngs him, forces him over, and he is at once forgotten, without even a half-suppressed sigh for his funeral oration. *Help yourself!* Is the watchword. The life of the genuine American is the soldier's life; like the soldier he is encamped, and that, in a flying camp, here to-day, fifteen hundred miles off in a month. ... Like the soldier, the American of the West take for his motto, *Victory or death!* But to him, victory is to make money, to get the dollars, to make a fortune out of nothing, to buy lots at Chicago, Cleveland, or St. Louis, and sell them a year afterward at an advance of 1000 per cent; to carry cotton to New Orleans when it is worth 20 cents a pound. So much the worse for the conquered; so much the worse for those who perish in the steamboats!
>
> The essential point is not to save some individuals or even some hundreds; but, in respect to steamers, that they should be numerous; staunch or not, well commanded or not, it matters little, if they move at a rapid rate, and are navigated at little expense. The circulation of steamboats is as necessary to the West, as that of the blood is to the human system. The West will beware of checking and fettering it by regulations and restrictions of any sort. The time is not yet come, but it will come hereafter [58, pp. 23-25].

The Franklin Institute Reports—
A Reasoned Technical Response to Catastrophe

In response to the tragedy of the *Helen M'Gregor* and the inability to pass steamboat safety legislation in the House of Representatives, the then Secretary of the Treasury, Daniel Ingham, was instructed to collect information and submit it to a House Select Committee. The report was given to the Congress two years later, May 18, 1832, in *Report 478*. At the same time, however, Ingham initiated a parallel independent investigation by scientists at the Franklin Institute of Philadelphia, the first federally funded scientific investigation.

Various explanations are given for Ingham and the Institute to cross paths. Both independently produced circulars soliciting information on boiler-explosions at the same time and for the same audience of boiler manufacturers and operators. Both used the same special observers in the East and West for detailed information; and Mr. Ingham himself lived just outside of Philadelphia in Bucks County [60, p. 6]. Correspondence between the Institute and Ingham resulted in the submission of a request from the Institute for $1500 to underwrite a series of experiments. The sum at the time was small since none of the experimenters was to draw a salary: all were volunteers, and the sum was only for the purchase of experimental equipment, sample boilers, and the work of toolmakers. The plan of research was approved by Ingham in October 1830.

In the way of written reports, the Secretary of the Treasury got his money's worth because four different reports were written, multiple copies printed, and the findings sent on to the House and Senate. Once the reports were sent on to the Secretary of the Treasury and Congress, they were serially reprinted in the *Journal of the Franklin Institute*.

For their part, the Institute and its experts wanted to participate in this research because they worried about the track record of Congress in carefully sorting out fact from fancy in boiler-explosions:

> Could the question meet with that thorough discussion & investigation before a committee of Congress which its importance & difficulty requires we should forbear recommending to the institute to take any steps in it; but it is well known that from the many subjects which press upon the attention of Congress, & from the indisposition to follow the British precedent of instituting enquires founded upon the evidences of practical and expert men, it becomes a difficult task for a committee of their body to possess themselves of a knowledge of the facts that have been developed during an experience of twenty years over a country so extensive as our [59, pp. 33-35].

Three of the Institute reports culminated in the *General Report* characterized as "cast in near-perfect form to secure legislative actions" [60, p. 19]. (All of these reports are available in microfiche and facsimile as listed in the endnotes.)

The first report, *Communications Received by the Committee of the Franklin Institute on the Explosions of Steam Boilers* (1832) [61, Fiche 148-149] came out about the same time as did *Report 478* from Wickliffe's House Committee, and it was very similar in purpose and organization.

The second report, *Report of the Committee of the Franklin Institute of the State of Pennsylvania for the Promotion of the Mechanic Arts, on the Explosions of Steam-Boilers, Part 1, Containing the First Report of Experiments Made by the Committee for the Treasury Department of the U. States* (1836) [61, Fiche 149-151] and the third report, *Report of the Committee of the Franklin Institute of the State of Pennsylvania for the Promotion of the Mechanic Arts, on the Explosions of Steam-Boilers Made at the Request of the Treasury Department of the United States, Part II, Containing the Report of the Sub-Committee to Whom Was Referred the Examination of the Strength of Materials Employed in the Construction of Steam Boilers* (1837) [61, Fiche 151-153] were both available to the House Select Committee on Steam Boiler Explosions that had been constituted in December 1836. These two reports in their technical complexity and quantity of raw data were unlike any other report created by the Congress in their steam boiler investigations of over a decade.

Finally, the *General Report*, or *General Report on the Explosions of Steam-Boilers by a Committee of the Franklin Institute of the State of Pennsylvania for the Promotion of the Mechanic Arts* (1837) [60, 61] was available in the summer of 1837 and 500 copies were made and distributed by the Secretary of the Treasury. What was most remarkable about this final and fourth report is that it was unlike any of the prior three and unlike any prior Congressional report in that it summarized in an easy-to-follow fashion the scientific information needed to frame effective legislation. The *General Report* even contained a ready-to-debate Congressional bill as an appendix so that theory, practice, and law were joined in a single report.

It is important to remember the sequence in which the reports were published and received by Congress because earlier data and material affect how readers read later data and material (see Table 1) [62, pp. 14-17]. For by the time the *General Report* was issued with its clear connection between experiments, research, and law, Congress had had to assimilate approximately 350 pages of reports [63] which had studiously avoided making connections between experiments, research, and law and had instead presented disconnected contradictory paradigms and findings. Moreover, just as the *General Report* was presented and perhaps began to be digested, the longest and scientifically most arcane report, *Part II on the Strength of Materials*, was delivered, and it too studiously avoided making recommendations or drawing conclusions. At the same time, the city of Charleston produced their report on the *Home* steamboat sinking and had sent it to every member of Congress—especially the members of their delegation in the House and Senate Select Committees. Given the constraints of time, the huge amount of data available, and the very limited technical and scientific background of the Senators at the time who came almost universally from lawyer backgrounds, one could question whether the *General Report* was ever successfully assimilated by

Table 1. Sequencing of Reports

Report Name and Date	Number of Pages	Organizational Principle
Report 478 issued by the Wickcliffe Committee (1832) [57]	200	**None**; it endeavored to be rhetorically free of intentionally favoring any information. Information was largely grouped by source.
Communications issued by the Franklin Institute (1832) [61]	69	**None**; it endeavored to be rhetorically free of intentionally favoring any information. Information was largely grouped in chronological order or by common focus on an explosion.
Part I. Explosions of Steam-Boilers issued by the Institute (Summer 1836) [61]	84	Began with an overview of the 12 areas of investigation that were originally proposed to the Treasury Secretary in 1830, described the experimental equipment in general and in specific, and then gave the results in tables. No general recommendations or conclusions were offered.
General Report issued by the Institute (December 17, 1836) [61]	48	A rhetorical masterpiece following both the letter and spirit of a classical disposition including: overview (*exordium*), overview of the causes of explosions into five areas (*divisio*), summary discussion of the relevant evidence related to the cause of explosion drawing on by reference *Report 478, Communications*, and *Part I*, including a refutation concerning promotion of safety if an explosion were to occur, and concluding with a summary and conclusion along with an appended Bill to move the audience to relevant action.
Part II. Strength of Materials issued by the Institute (January 7, 1837) [61]	247	Began with an overview of the principal (9), incidental (6), and subsidiary (6) areas of investigation, described the experimental equipment in general and in specific, and then gave the results in tables. No general recommendations or conclusions are offered. Included are 110 tables (some proceeding over two pages), many graphs and figures, and an index.

the time of the vote on S. 1. In fact, when Walter Johnson wrote to the Secretary of Treasury in June of 1835 to explain why his subcommittee's report on the strength of materials was delayed, he noted:

> I cannot see in the disasters which have recently occurred any evidence that a want of information was the source of the evils, but on the contrary the strongest indications both in the case on the Hudson & that at Memphis of recklessness which would alike disregard the deductions of science & the counsels of prudence until public opinion should have effected a complete abolition of the practice of placing incompetent and unprincipled persons in stations to control the machinery of boats and to sport with the lives of their fellow beings [64, pp. 93–113].

In a time when technical writing was not a subject to be learned from books or teachers, it is most instructive to see how the members of the Franklin Institute moved from a rhetoric of technical reports that was characteristic of the time to a rhetoric of technical reports that broke new ground. Moreover, it is quite interesting to watch the rhetoric of two competitors on the Institute Committee, Alexander Dallas Bache—primary author of the *General Report*, and Walter Johnson—primary author of the *Part II Strength of Materials Report*.

Bache's report concentrated on making the technical information of the investigations accessible to non-expert members of Congress, while Johnson's report seemed to concentrate on the cutting-edge research information regardless of whether the non-expert members of Congress could understand it. Both used the reputations gained in their reports and research for attaining important posts in the world of U.S. antebellum science and technology: Bache went on to a variety of important government posts including the appointment as head of the U. S. Coast Survey [65], and Johnson went on to various research projects for the Secretary of the Navy and various Congressional committees [66]. One can watch these two authors grow and develop in their rhetoric because both left behind a trail of reports that can be followed quite easily.

Traditional Technical Writing of the Era—Communications Received by the Committee of the Franklin Institute on the Explosions of Steam Boilers (1832)

Report 478 given to the Congress on May 18, 1832 was massive [57]. Much of this material was elicited by the preparation of interrogatories sent out by the Secretary of the Treasury in circulars across the country [67]. *Communications* by the Franklin Institute, though not quite as massive—only 69 pages—was also the result of information solicited by circular and also contained the similar mélange of information [68]. The two 1832 reports by the Treasury and the Institute were also similar in that they imposed no critical overall organization on the materials beyond vague grouping, in the Treasury's effort, and chronological order of reception in the Institute effort. Both were very much like the pre-paradigmatic random fact-gathering described by Thomas Kuhn in his *Structure of Scientific Revolutions*:

> In the absence of a paradigm or some candidate for paradigm, all of the facts that could possibly pertain to the development of a given science are likely to seem equally relevant. As a result, early fact-gathering is a far more nearly random activity than the one that subsequent scientific development makes familiar [69, p. 15].

For example, here's how Louis McLane, then Secretary of the Treasury, described in *Report 478* how the information was published:

> 1st, Returns received from the several [duty] collectors, marked A, and numbered 1 to 58 inclusive. 2nd, Various communications in answer to "interrogatories in relations to bursting of steam boilers," which are marked B, and numbered 1 to 31, inclusive. 3rd, Translations of circulars, marked C, and numbered 1 and 2, from the Director General of roads, bridges, and mines, in France to the Prefects of the departments; being all the information in possession of the department. . . . publishing a notice in various newspapers throughout the United States, and issuing the circular to collectors, of 31st October 1831, copies whereof, marked D and E, are herewith transmitted [57, p. 9].

Moreover, the House Committee made a point of rhetorical non-interference by noting that the collection was devoid "of the expression of any opinion of their own." Most of the material was simply numbered and grouped by source (e.g., U.S. port toll collectors or French Prefects). Although there was some mention in *Report 478* of placing

> . . . all the papers, preparatory to submitting them to the House, in the hands of one or more gentlemen, competent, from their knowledge of the subject, to pronounce upon the practical value of the facts and suggestions they present, with the hope that, by the aid of the information contained in them, and that which the person selected might add from his own and other sources, some scheme of measures might be suggested for promoting the object for which the inquiry was instituted [57, p. 9].

This scheme of organization was never enacted in the published report. Then there was a paragraph describing the Franklin investigations and how they were as yet incomplete. Perhaps in this little introduction, there is a suggestion of what was the original scheme for which the Secretary of the Treasury pursued a two-pronged investigation using the Institute—the experts at the Institute were supposed to function in the capacity of wise referees of the information collected. But it never happened.

The introduction to the Institute's *Communications* also contained a description of their intentionally weak method of organization that would avoid any sense of partiality:

To avoid any appearance of partiality, the committee have determined to publish these letters in the order of their date, but they have been induced occasionally to deviate from this rule, so as to group together such as refer to the same accident. By exhibiting the views of different writers upon the same explosion, it will be easier to come to the truth as to its real cause. Thus, for example, in relation to the *Helen M'Gregor* and *Caledonia*, the committee will have the pleasure of presenting in connection, several letters which mutually confirm each other [70, pp. 1-2].

Thus, in 1832 in these two reports written by different groups, there was simply a rather random grouping of facts as both the Treasury and Institute investigators groped for a paradigm to organize the material. Moreover, the *Communications* seemed to use two separate paradigms to explain the explosions. One was that it is the problem of the technology itself—the technology was inadequate to contain the pressures created, the safety valves needed to be more numerous and better constructed, and other safety devices were needed. The interrogatories sent out by the Institute to elicit such data implicitly directed the respondents to focus on such technical areas:

These accidents ought not by any means to be considered as an unavoidable consequence of steam navigation. They proceed, it is believed, in most cases, from defective machinery, improper arrangement or disposition of the parts, and finally, from carelessness in its management. That the causes of accidents may be partially, if not wholly, removed by salutary regulations, appears highly probably . . . [68, pp. 34-35].

Some respondents, however, focused on the need for better training because the technology was so novel:

The introduction of steam navigation has been so rapid that men of character, capability, intelligence, skill and sobriety, were difficult to be procured in sufficient numbers . . . [68, p. 44].

while a few focused on areas that civil liability punishments could deal with:

In the first place I assert, without fear of contradiction, that two-thirds of the accidents, are to be attributed to nothing but ignorance; and from this cause, that men who are perfectly acquainted with their business cannot be had except at high wages; and though a mistaken notion of economy, a fireman is taken, who has gained the experience of stopping and starting an engine, but who is entirely ignorant of its principle; aye, some who could not tell you whether the engine they were attending was high or low pressure! And thus the grandest machine ever invented, and the lives of thousands are placed in the hands of ignorance. No wonder that explosions occur! [68, pp. 46-47].

Report 478 was similarly split between these paradigms having letters that asserted on the one hand:

> Steam is an element or principle completely within the control of science, and that every accident that has occurred has originated from want of skill, or want of judgement and care; that the accidents that have occurred, in most cases, have been where engineers of experience have been replaced by ignorant or careless individuals, who are obtained for a less compensation [57, p. 34].

And, on the other:

> In fact, when it is considered how many causes may be assigned, embracing the wide range of competition in speed, neglect of engineer and firemen, imperfect metal, &c., &c., I come irresistibly to the conclusion, that, though care and skill has done much, and may do more toward preventing accidents of this nature, yet accidents are likely to recur, and the wisest course is to devise some plan to preserve life at least, till experience and practice develop a certain remedy against explosion of boilers. With this view, I proceed to offer a cheap and practical safeguard [57, p. 56].

In the latter cases

> Explosions which take place . . . may truly be said to be beyond the preventive power or control of the engineer. He cannot tell, when called to the management of an engine on board of a steamboat, that there has been a fault in the construction of the boilers—a defect in the material out of which they were composed—or that, by its too long use, the original strength has been so far impaired as not to be capable of sustaining the ordinary pressure of steam which belongs to the capacity of the boiler. He may not know how long the boilers or the boat have been in use; consequently, no skill of his, thus situated, is, or can be, competent to guard against explosions produced by any or all of these causes [57, p. 3].

While in the earlier cases

> It only belongs to legislation to excite, by rewards and punishments, that faithful application of those engaged in its use, which will best guard against the dangers incident to negligence [57, p. 2].

Not only were the paradigms of causation and solution similar in both the Institute's *Communications* and *Report 478* , but at least four of the letters in the Institute's *Communication* were duplicated in *Report 478*. Moreover, although the Secretary of the Treasury was unable to get the experts of the Institute to review the answers submitted and put them in some order, the largest message included in *Report 478* came from the leader of the Institute's sub-committee investigating the strength of materials, Walter Johnson.

It is finally interesting to note that although both the *Communications* and *Report 478* were published at about the same time and focused very directly on steamboiler explosions aboard steamboats, the proposed Bill offered by the Wickliffe Committee on pages 6 to 9 of *Report 478* appeared to respond to something other than the material in the other 197 pages of the report. Three of the twelve sections in the draft legislation, 25 percent of the items in the bill, focused on lifeboats, fire-fighting issues, and safe passing at night—items of which were never discussed in the actual report itself. Thus, there was evidence in *Report 478* itself that the politicians ignored most of the specific technical solutions and over-emphasized the civil liability solutions, as well as responding more to pressures and interests outside of the report than the data contained within.

Report of the Committee of the Franklin Institute of the State of Pennsylvania for the Promotion of the Mechanic Arts, on the Explosions of Steam-Boilers, Part I, Containing the First Report of Experiments Made by the Committee for the Treasury Department of the U. States (1836)

Four years after the 270 pages of the preliminary collections of reports and letters in the Institutes's *Communications* and *Report 478*, the first of the Institute Reports was published, and distributed throughout Congress. *Part I* had as its object

> To test the truth or falsity of the various causes assigned for the explosions of steam-boilers, with a view to the remedies either proposed, or which may be consequent upon the result of the investigation [71, p. 3].

and it was organized along the lines of a traditional scientific report:

> The committee propose first to give a general description of the apparatus used, followed by details in the more complex parts; next to report the results of their examination upon each of the questions proposed for investigation [71, p. 3].

The items the Institute investigators looked at were twelve in number and repeated the list of experiments originally outlined to Secretary Ingham in 1830 when he had first approved the grant of federal money to the Institute [72]. Thus, with the delivery of the *Part I* report, the Institute had delivered precisely what it had promised. Moreover, it is interesting to note how the order of presentation in *Part I* changed in comparison with the original 1830 proposal—these differences are noted in brackets:

1. Examine the efficacy of gauge-cocks, glass gauge, and a newly invented gauge-cock when foaming in the boiler may occur or other commotion. **[Was not even proposed in 1830]**
2. Examine whether highly heated metal can suddenly produce large amounts of steam thus replicating the experiments by Klaproth. **[1st item in 1830 proposal]**
3. Examine whether the introduction of water into steam decreases or increases the amount of steam. **[2nd item in 1830 proposal]**
4. Examine whether steam in contact with water changes it properties. **[3rd item in 1830 proposal]**

5. Examine the efficacy of fusible metal plates to prevent explosions. [**Was not even proposed in 1830**]
6. Repeat Klaproth's experiments with copper and iron under different circumstances and quantities of water. [**6th item in 1830 proposal**]
7. Examine to see if any permanently elastic fluids are produced by highly heated metals. [**5th item in 1830 proposal**]
8. Accurately observe boiler bursting caused by gradual increases of pressure. [**4th item in 1830 proposal**]
9. Examine if a repulsion exists between steam and super heated iron replicating the experiments of Perkins. [**8th item in 1830 proposal**]
10. Examine whether a safety valve can remain stationary, untripped, by steam reaching pressures which are suppose to trip it. [**9th item in 1830 proposal**]
11. Examine the effects of sediment deposits in boilers. [**10th item in 1830 proposal**]
12. Examine the relation between temperature and pressure at ordinary working pressures. [**Was not even proposed in 1830**]

Although *Part I* was a rhetorical step forward in presenting evidence about explosions in an informed way, it was still very much of a grab bag. For example, the problem with the effectiveness of boiler instrumentation (gauge-cocks, fusible plates, and safety valves) was spread out over items 1, 5, and 10. The relation of superheated metal and the rapid production of steam was spread out over items 2, 3, 6, and 7. There also appeared to be some basic science investigations in items 9 and 12.

Thus the organizational principles behind *Part I* seemed less related to enhancing the easy application of its findings for remedies and laws and more to prove that the Institute had fulfilled its contractual obligations to the Treasury Department. This is rather directly suggested by the following statement in the report:

> The Committee of the Franklin Institute on the Explosions of Steam-boilers, respectfully present to the Secretary of the Treasury, their report of the experiments undertaken at the request of the department. The queries which were submitted by the committee, to the officer at whose request the experiments were instituted (Secretary Ingham) have formed the basis of the labours of the committee.

Thus the Committee had their immediate audience in mind—the Secretary of the Treasury—but missed designing *Part I* with their ultimate, decision-making audience in mind, the non-technically-oriented Congressmen who would frame a bill, and the engineers and fireman who would use the research in daily application.

Part I did, however, establish a much more practical bent to the research than that originally proposed by the Committee in 1830. For example, in the original proposal much emphasis was laid upon the replication of Klaproth's and Perkins's experiments—an interest of a more "basic" research bent. And, indeed, the largest number of pages in the report had to do with these experiments. Moreover, this was the material that their fellow scientists and mechanicians cited in their own subsequent reports and treatises. However, when *Part I* was published, it included items 1 and 5 that specifically dealt with practical instrumentation and specific safety items such as fusible plates or plugs. (Safety devices allowing pressure above a specified level to intentionally blow out of a boiler. These plates or plugs would blow out before the whole boiler would because the plugs were made of an alloy which melted at lower temperatures than did cast or wrought iron.)

General Report on the Explosions of Steam-Boilers by a Committee of the Franklin Institute of the State of Pennsylvania for the Promotion of the Mechanic Arts (1837)

When Bache (see Figure 22) came to write the *General Report*, he indeed designed it to be perfectly cast for usefulness in the Congress. The *General Report* was actually superfluous—the Institute had fulfilled its obligation to Congress completely with the delivery of *Part 1*. Yet what the *General Report* did was to

1. combine research from all previous reports (*Report 478, Communications, Part I*, all subsequent articles appearing in the *Journal of the Franklin Institute,* and even legislation (e.g., the 1834 draft legislation issuing from Southard's Committee on Naval Affairs), and
2. use the classical five-part organization of Ciceronian dispositio [73, pp. 63–89].

There was no table of contents nor index, and yet the 42 core pages of the *General Report* could be easily read at one sitting through the frequent use of headings, numbered items, and overviews.

The *General Report*'s introduction quickly established the problem— "a series of disasters by which human life is sacrificed called loudly for an investigation" [60, p. 3]—and hoped to allay the apprehensions of the public that "accidents were not unavoidable incident to the useful agent they distrusted" [60, p. 3]. From the scientific viewpoint, the *General Report* also hoped not only to explain the causes of the explosions but also to explain what were not causes so as to "turn away the attention of ingenious men from false hypotheses" [60, p. 4]. And whereas all earlier reports had on a stance of studious impartiality, the *General Report* explicitly *selected* the most pertinent points and omitted others [60, p. 5].

The *General Report* quickly moved to an overview of how the rest of the report would proceed, the Ciceronian partition:

Figure 22. Alexander Dallas Bache, primary author of the *General Report.*
The Historical Society of Pennsylvania (HSP), Alexander Dallas Bache,
Newsam Collection, P. S. Duval, Lithographer, Philadelphia.

The Committee propose to examine separately the circumstances which they
consider as the proximate causes of explosions in steam-boilers, and the
preventives or remedies which have been proposed to meet them. Under each
division of the subject they will make the suggestions or recommendations to
constructors and others, which they base upon the previous discussion; and at
the close of the Report, will present a project of a law for carrying into effect,
in regard to steam-boat boilers, those recommendations which are of primary
importance [60, p. 5].

The most dramatic illustration of Bache's rhetorical expertise in the *General Report* was in the sequencing of the *conformation*, the core arguments of the report. This portion of the *Part I* report was based upon the rather random list of twelve items in the original proposal to the Secretary of Treasury; evidently the authors did not feel they had much room to re-work and re-order the parts. But in the *General Report,* these 12 items which might overwhelm a non-technical audience, were re-combined into five problems and solutions:

1. Explosions from undue pressure within a boiler, the pressure being gradually increased
2. Explosions produced by the presence of unduly heated metal within a steam-boiler
3. Explosions arising from defects in the construction of the boiler or its appendages
4. Explosions resulting from carelessness or ignorance of those intrusted with the management of the steam-engine
5. An examination of the particular cases of collapse of a boiler, or its flue by rarefaction within [60, p. 6].

The second cause combined the separate investigations in *Part I* having to do with the problem of deposits or sediments, the investigations of copper and iron, replicating Klaproth's experiments, the argument against Perkin's hypothesis concerning saturated steam and the introduction of water, along with specific examples of explosions. Each paragraph in the *General Report* was numbered for easy reference in Congressional discussions, and four solutions were examined. Each section then summarized an operational recommendation for engineers and firemen.

In classical rhetoric, once the main argument had been given, a *refutation* voiced the words of a critic and then responded. With a single page left in the central document, the *General Report* considered how one could increase safety to passengers if—even with all remedies used and full understanding applied from the *General Report*—an explosion were to take place. And here, the Committee urged the construction of a strong bulwark between the engine and the rest of the ship.

Finally, the peroratio in a Ciceronian arrangement summarized the main points of the arguments and moved the audience to action—and certainly this was the intention of the *General Report* when it included draft legislation with 20 sections including a Sec. 13:

> And be it further enacted, That whenever the mast of any boat or vessel, or the person or persons, charged with the navigating said boat or vessel which is propelled in the whole or in part by steam, shall stop the motion or headway, of said boat or vessel; or the said boat, or vessel, shall be stopped for the purpose of discharging, or taking in cargo, fuel, or passengers; he, or they, shall keep the engine of said boat, or vessel, in motion sufficient to work the pump, and give the necessary supply of water, under the penalty of _____ dollars for each and every offence in neglecting or violating the requirements of this section.

Here, in its very strength of connecting disinterested science and public legislation, was where readers' could begin to suspect that the *General Report* was less a disinterested summary of the science of steamboat explosions, than a Trojan horse for Whig policy. For example, it might have seemed suspicious to Democrat readers and especially to conservative Democrat nullifiers such as Calhoun that 12 of the 20 separate sections were taken almost verbatim from Whig Samuel Southard's Naval Affairs Committee legislation of 1834, and that even parts of the *General Report*'s draft legislation not taken directly from Southard's bill were taken from the speech given in the Senate by Whig leader, Daniel Webster. For example, the draft legislation offered in the *General Report* had this statement

> Said gauge and scale shall be so placed as to be readily examined by any and every passenger on board of said boat.

Webster had said nearly the same thing on the floor of the Senate in 1833:

> a proper apparatus to show at all times the level of the water, and also the intensity or height of the steam, and this apparatus so arranged that its indications may be seen from without, . . .

This provision of the steamboat law did not appear in any other draft of the legislation. Moreover there were some civil liability provisions which were included in the *General Report*'s draft legislation which were never discussed in any of the Institute's reports. Thus the very unassailable disinterestedness of science which was the primary appeal of the Institute to Congress and the Secretary of Treasury might have been undermined in the eyes of partisan readers by how the *General Report* explicitly became partisan in choosing and omitting certain legislative provision, and in using the Naval Affairs Committee's draft legislation as the vehicle to contain their own unique sections (see Table 2).

Report of the Committee of the Franklin Institute of the State of Pennsylvania for the Promotion of the Mechanic Arts, on the Explosions of Steam-Boilers Made at the Request of the Treasury Department of the United States, Part II, Containing the Report of the Sub-Committee to Whom Was Referred the Examination of the Strength of Materials Employed in the Construction of Steam Boilers (1837)

Johnson was 42 when he presented *Part II*, Bache was 30 when he presented the *General Report*. Johnson was Harvard educated, lectured to mechanics at the Institute, and drew his primary salary teaching at the Institute's High School [66, pp. 93-113]. Bache entered West Point—the premier engineering institution in the United States at that time—at the age of 15, and, once graduated, became one of its youngest instructors. He moved to Philadelphia where he accepted a professorship at the University of Pennsylvania. The *Strength of Materials* investigation was

Johnson's first major investigation, while Bache had preceded his Chairmanship of the steam-boiler project with investigations into water-wheels, weights and measures, and meteorology. Bear these differences in mind because the rhetoric and approach of Johnson and Bache clearly put them at odds with each other, caused friction that was determined by an ultimatum being issued by the Institute to Johnson to finish his work, and could only be worked out by making Bache and Johnson each Chair of their particular subcommittees.

Walter Johnson (see Figure 23) first made his written appearance in the steamboat reports in the longest report in *Report 478*, an account of 42 pages of some 173 different experiments with most of the results presented in 16 tables, some of which extend over two and three pages. Primarily, Johnson was doing "basic" research in attempting to develop a law of "action between a heated surface and water at different temperatures" [57, p. 125], but he introduced this only generally in the opening to his section of *Report 478* where he stated rather generally that he wanted to find out the relationships between

1. Quantity of steam produced;
2. Weight of material;
3. Surface exposed;
4. Time of action; and
5. Period of greatest effect.

Johnson nicely applied his experimental results initiated in a very general way to six specific points:

1. *The temperature of most rapid vaporization*;
2. *The name of the phenomena exhibited at that point and immediately above and below it*;
3. *Effects of lubricating the surface*;
4. *Influence of mechanical pressure to bring liquid in contact with hot metal*;
5. *Action of hot metals on other liquids*; and
6. *Nature of repulsion and degree to which heat is transferred from metal to liquid.*

Yet Johnson's conclusions appeared to have little application to legislation and practice, and thus his section was unique in a report which primarily conveyed very practical observation and opinion. Additionally, he weakened the ability of readers to connect his basic research to steamboat legislation by concluding with a long description of his experimental apparatus.

This example of Johnson's rhetorical approach to report writing in *Report 478* is quite similar to his approach to *Part II* with terrific "basic" research findings and an extremely poor effort at making such findings applicable to legislation or practice. *Part II*, for example, is the only one of the Institute's reports that was not designed to be read from front-to-back: it has 247 pages (longest of all the reports), 102 tables (some over two and three pages), along with numerous graphs and

Table 2. Comparing the General Report's Provisions to Previous Legislation and to Grundy's Proposed Bill

1836—Franklin Institute Report's Proposed Law [59, pp. 44-48]	1824 Vinton's Proposed Law [74]	1830–2 Wickliffe's Proposed Law [57]	1833 Webster's Proposed Law [75]	1834 Naval Affairs Committee Proposed Law [76]	1838 Grundy's Law Proposed [77, p. 8; 24, Appendix B]
Technical and Operational Solutions					
Boiler and machinery inspections	X	X	X	X	X
Certificate of boiler pressure limits	X	X	X	X	X
Two safety-valves in place	X	—	—	—	—
One of two safety-valves locked	X	—	X	—	—
Pressure gauge included	—	—	X	—	—
Fusible metal safeguard alarm	—	—	—	—	—
Not clearing sedimentary matter buildup out of boilers is punishable by manslaughter	—	—	—	—	—
No part of boiler is directly exposed to flame without contact with water	—	—	—	—	—

					Technical and Operational Solutions Not in Franklin Report
Inspectors shall test boiler pressures	X	—	X	X	X
Periodic testing shall take place	—	X	—	X	X
When boat is stopped, machinery will be kept in gear so that water pump is kept activated	—	Safety valve opened; X	—	X	Safety valve opened; X
Machinery shall be controlled by practical mechanic: 21 years old, experienced, with testimonials	—	—	—	—	—
There shall be a bulwark between the machine and passenger compartments	—	—	—	—	—
Must carry two long boats		X		X	X

Table 2. (Cont'd.)

	1836—Franklin Institute Report's Proposed Law [59, pp. 44-48]	1824 Vinton's Proposed Law [74]	1830-2 Wickliffe's Proposed Law [57]	1833 Webster's Proposed Law [75]	1834 Naval Affairs Committee Proposed Law [76]	1838 Grundy's Law Proposed [77, p. 8; 24, Appendix B]
Must shut off steam when passing an ascending vessel at night—not in final bill		—	X	—	X	X
Must carry suction hose and fire hose		—	X	—	X	X
Night running requires signal lights		—	X	—	X	X

Civil Solutions

Licensing of all steamboats	X	X	X	X	X
Tampering with safeguards is punishable	X	—	—	—	—
If Captain, Master, or Engineer is drunk, racing, gambling, shall be wholly liable for death, injuries, and damage without any insurance	—	—	X	X (drunken-ness not included)	X
If Captain, Master, or Engineer is drunk, racing, gambling, and there are deaths, injuries, or damage there is a mandatory sentence and fine	—	—	X	X (drunken-ness not included)	X
False certificates or licenses are liable to penalties	—	—	—	—	—

Figure 23. Professor Walter Johnson, primary author of the *Strength of Materials* Report. *The Historical Society of Pennsylvania* (HSP), Walter Rogers Johnson, Gratz Collection, John Sartain, Engraver, Philadelphia, C.7 B.23.

figures. Indeed, *Part II* was supposed to be selectively read like a handbook because it alone of all the reports had a table of contents and an index.

Johnson began with an overview that suggested he would be looking at "Principal," "Incidental," and "Subsidiary" questions, but observed, quite correctly, that

> The discussion of the questions above enumerated, will necessarily follow an order somewhat different from that in which they are here stated [61, Fiche 151-153].

In fact, the investigation to the first "Principal" question concerning rolled boiler iron did not appear until page 79 and only after the experimental apparatus had been exhaustively described as well as the origin and preparation of the materials and the answer to "Principal" question 2 concerning rolled copper. Thus, no matter what order Johnson announced was in his report, he reported in a wholly other order.

Johnson's report and his experiments basically invented the science of materials strength testing, and his findings were to form the base of many others' inventions and improvements [78, pp. 34-37]. Yet as the historian Bruce Sinclair observed,

> The form and style of his report was quite different from those of the others which issued from the investigation. There was no impetus for reform in Johnson's report, nor was it framed to elicit legislative actions [59, p. 12].

Contemporaneous Reactions to the Institute Reports in the Scientific Community: Hales's Open Letter to Grundy, Locke's Cincinnati Report, and Steam Textbooks by Renwick and Ward

The first contemporary response to The Institutes's *Part I* came in a May 1838 open letter to Felix Grundy by a William H. Hale that was published in the *Journal of the American Institute* [79]. In his five-page letter, Hale quotes from the *Journal of the Franklin Institute*'s 1836 serialized version of the *Part I* seven different times to buttress Hale's own points and arguments, i.e., "This is within my own experience, and is also proved by the experiments of 'the Committee of the Franklin Institute'" [79, p. 531]; or "This theory of explosions is fully supported by the report of 'the Committee of the Franklin Institute.'" Hale does criticize the Institute's finding that explosions can happen by a gradual increase in pressure and not just by dramatic increases in pressure, but his general attitude is that the information in *Part I* is true and accurate. Hale was matched in this attitude by a larger Committee appointed by the American Institute to investigate the causes of the explosions of steam-boilers because they directly quote one table from the *Part I* without commentary or critical evaluation [80, p. 649].

The second contemporary response came in a lengthy report written that same summer in June 1838 [81, p. 355; 82, pp. 21-23]. Dr. Locke, a professor of chemistry at the Medical College of Ohio, was the primary author of the *Report of the Committee Appointed by the Citizens of Cincinnati to Enquire into the Causes of the Explosion of the Moselle* [83]. In his 1838 report, after Locke had reconstructed the *Moselle* explosion, he began offering a background to explosions of high pressure steam-boilers on pages 29 to 44, and it was in these pages that Locke specifically referred to the "very able" Institute's reports 13 different times, and used them as his experimental foundation. (It is interesting to note that the very scientific Locke perused over both the *General Report* and *Part I*, but did not once mention material from *Part II* even though it had been given to Congress and published serially in the *Journal of the Franklin Institute* for over a year.) But since Locke saw as his goal "a piece of brief instruction to plain, practical men," he excerpted and used a simpler approach yielding such aphorisms as his own:

> The expansive force of steam produced in a closed boiler contains a sufficient supply in water increases as the heat increases, but not at the same rate, the force increasing faster than the heat [83, p. 29].

At other times Locke would directly quote *Part I* for other aphorisms that would be understood by "plain, practical men":

> All the circumstances attending the most violent explosions may occur without a sudden increase of pressure within the boiler [83, p. 39; 71, p. 69].

Locke's observations on sediment and their effects on boilers (Point 11 in *Part I*) were also quoted directly from *Part I*. However, Locke re-organized the Institute's information by first exploring the causes of explosions and then the remedies for explosions whereas *Part I* began with the gauges and remedies and then proceeded to the causes of explosions.

The majority of references were to *Part I*, which Locke used as his "basic" research for making general observations and offering solutions—a number of which he drew from the *General Report*, such as the use of a green glass tube rather than gauge cocks to indicate water level in the boilers and the need for thermometers. Locke used *Part I* primarily to explore the evidence of foaming and how that interfered with gauges as well as refuting Perkins's idea that injection of water will always increase pressure [83]. Also Locke looked to *Part I* to reveal that water injected into a hot boiler would not decompose into such explosive gases as hydrogen, another Perkins theory.

Another way to gauge the scientific community's reaction to the Institute's reports could be seen another year after Locke produced his report. Professor James Renwick of Columbia produced the first edition of his *Treatise on the Steam Engine* in 1830, six years **before** the Institute reports, and Renwick's second edition was released in June of 1839, one year **after** the reports were published [85]. Renwick's chapter on boilers directly focused on the matter considered in the Institute reports. As a result, in the Second Edition Renwick made 18 additions and deletions, and over half contained material from the Institute's reports that had been assimilated or questioned (see Table 3) [86, p. 85].

In the end, *Part I* and *The General Report* provided valuable information which could eventually lead to remedies and which could be understood and applied.

One other place to look for reactions in the scientific community to the Institute's reports could be in the military. Captain James H. Ward wrote a 1845 book in the Philadelphia Shipyard Naval School on ordnance, gunnery, and steam for the instruction of midshipman. He subsequently left the military or was somehow able to publish the second half of his book, the part on steam that would be applicable to civilian interests, under the title of *Steam for the Million. An Elementary Outline Treatise on the Nature and Management of Steam* [87.] When in the course of discussing the Perkin's theory of saturated and surcharged steam, Ward wrote:

> Elaborate experiments made by the Franklin Institute of Philadelphia in 1836 show, however, with apparent conclusiveness, that no such effect results; but that on the contrary a small reduction of elastic force is produced by the injection of water to surcharged steam [87, p. 12].

By 1845, the Institute's Reports, at least to the scientific community, were conclusive.

Contemporaneous Reactions to Institute's Reports by Those Most Directly Involved: Steamboat Inspectors, Engineers, and Firemen

It quickly became apparent that those most directly concerned with steamboat power did not understand the *General Report*. For example, one Jacob Walter, a steamboat inspector, wrote in a Louisville paper (and had it reprinted as *House Document 173* in 1839) the following observation regarding the state of understanding steam **after** the *General Report*:

> Many have been theorizing in a speculative way on steam, and the subject will be continued until truth shall be fully established; but as yet I have never seen a proper definition of the chemical department of steam engineering, or the process of generating steam fully or rationally explained, in a full and comprehensive manner, with its attendant phenomena. The great principle, heat, being but indifferently understood by the most scientific part of mankind, is one cause of the difficulty; and the great combination of science so necessarily involved in a subject like this make the difficulty still greater [88, p. 3].

This may have been just the words of a single person, even if he was an official steamboat inspector, but even more telling was the petition submitted to the House and Senate in 1843—nearly **seven years after** the *General Report* was published—by the Cincinnati Association of Steamboat Engineers which noted the *General Report* by name:

> In the following we find a discrepancy which we copy from the report of the Franklin Institute, which differs entirely from the results obtained by all former experimenters on the cause of steam boiler explosions. In a printed report of their committee, we find it stated that in twelve actual experiments they discovered that, by injecting water into a boiler containing highly heated steam, not entirely empty of water, and empty of water, and red hot, in every instance it diminished the temperature and pressure or expansion force of the steam in the boilers. This article refutes the doctrines of Mr. Perkins (page 71 of Allan's *Science of Mechanics*,) and contradicts the results of all other

Table 3. Renwick's Alteration in His *Treatise on the Steam Engine* in Response to the Institute's Reports

Second Edition Page	Material Altered in *Treatise on the Steam Engine*	Related to Institute Report
58	Tubular boilers create inconveniences	*General*
59	Klaproth's experiments replicated	*Part I*
60	There is deleted material from an earlier edition which stated that tubular boilers were free from the risk of explosion and this was replaced by noting that tubular only meant those carrying the heat, not tubes of water, and that these tubes carrying the heat were dangerous in how they were physically placed next to combustible material	—
62	Tenacity of wrought iron is increased by heat	—
68	The subject of forcing pumps and placements of gauge-cocks is turned from really relevant to primarily a historical piece since the low-pressure engines were rapidly diminishing	—
69-70	The subject of using a float as instrumentation in a moving engine is added with particular material from a Hall of Glasgow	*General*
71	Worried about problems with self-activated pumps	*General*
73	Safety valve openings for high pressure should be as large as for low pressure	—
76	The ratio of the arms of the safety valve lever to the surface of the valve should be 5:1	—
77	Douglas suggests placing a valve inwards to prevent the diminishment of space when it is taken up by steam	—

Page		
82	He qualifies his 1830 assertion that the decomposition of water into explosive gases is the "sole" cause of explosions to be "only causes"	disputed by *Part I and General*
83-84	The injection of water into steam quickly produces added steam and higher pressure (the Perkin's hypothesis) is disputed with the Institute's findings—this is the point that the Cincinnati engineers fasten on in their 1843 report on the confusion of findings	*Part I and General*
83	Tenacity of copper decreases with temperature while iron increases up to a certain temperature at which it too weakens	*General Report*
86	Addition of fusible plates as safety valves is added	*Part I*
88-89	When Renwick lists the nine things that ought to be done to prevent explosions, he adds the danger of internal flues (page 60 reprise); he backs off from advocating self-activated water feeding mechanisms (p. 71 reprise); he adds how foaming undercuts the valve of the gauge cocks; finally at the end of the list in 1830 when he noted how few of these safety devices were in general use, he burst out with "Need we wonder that explosions have become frequent, and that they have produced the most fatal consequences?" (p. 103). In 1839, his list is unchanged, but his outburst is deleted	—
90	Renwick injects an entire section condemning the use of steam chimneys in which the stacks out of which heat issues are wrapped in a water boiler. The added power is not worth the problems of safety	*General*
92	The need to "blow off" sediment and of cleansing and scraping boilers is advocated	*Part I and General*
96	He adds material on the boilers of locomotives	—

former experimenters. There were always dangers, says Mr. Perkins, to which boilers are exposed, against which safety valves present no security.

. . .

Among the various and respectable authorities referred to, who stand conspicuous among the scientific world for their deep research and unquestionable talents, who have investigated the nature and properties of steam, and the causes and effects of boiler explosions, we find that the conclusions they have arrived at are as various upon this interesting subject as the different writers vary in number. Their doctrines differ, also. One is based upon philosophical theory, another upon actual experiments, another upon visionary conjectures—all opposite and contradictory in their conclusions. Whether they arrived at them in a proper way or not, we are not accountable. This we do know, that the writers make this subject appear most vague, unaccountable, and mysterious to us; in fact nothing systematic can be comprehended by which we can be directed, in the management of our business, to ensure safety to ourselves or those entrusting their lives in our charge, if we had not a sure system of management, acquired by long practical experience and attentive observations [89, pp. 3–5].

It appeared that when presented with contradictory information, the Cincinnati Association of Steamboat Engineers did not feel they had the means to resolve the contradictions nor did the *General Report* persuasively convince them.

Yet in the end, in response to yet another resolution from the Senate to investigate the causes of steam boiler explosions, the Commissioner of Patents issued *Report 18* in 1848 [31] that was lengthy, 184 pages. In its first fifteen pages, the Institute's *General Report* was cited by Commissioner of Patents Burke 25 times, and he then reprinted a ten-page abstract of the *General Report* and a 70-page abstract of the *Part II* report. Burke was anxious to do this because, as he noted in the *Report 18,*

The labors of the Franklin Institute committee have been considered as the most valuable additions to the amount of our knowledge on the subject, and they have therefore been largely quoted. . . . But the undersigned is of opinion that, though the present report may throw no new light on the subject, yet the service of presenting, in one condensed general view, a resume of the present state of knowledge in relation to this subject, in a form intelligible to the general reader, may be of some value in its bearing upon intelligent legislation, and as affording a useful source of information to the practical man, perhaps deficient in scientific knowledge into whose hands this report may possibly fall [31, p. 31].

When the Department of Treasury originally received the *General Report* in 1836, it had 500 copies made. Now, twelve years later, in 1848, this same report was ordered reprinted by the Senate and Patent Office, and 10,000 copies were made.

Endnotes

1. To the parents of Carol and Arthur Wierum, of Brooklyn who perished on the *Metis* disaster, Words by George Dana, Music by Samuel Mitchell, Boston: G. D. Russell & Company, 1872. Oddly, the same steamboat disaster also produced two other songs, "Kiss me Mamma, for I am Going to Sleep" by J. A. Butterfield and "The Song of the *Metis*" by A. C. Atwood.

2. *RiverNews, Volume 2: Steamboat Disasters: Newspaper Accounts of Wrecks, Explosions, and Other Calamaties*, Compiled and edited by Milich Kouns and Ken Hulme, Steamboat Press, Salem, Oregon (1055 Fairview Ave. SE), 1999.

3. James T. Lloyd, *Lloyd's Steamboat Directory and Disasters on the Western Waters*, James Lloyd, Cincinnati, Ohio, 1856. (Facsimile reprint by Young & Klein, Inc., Cincinnati, Ohio, 1979.)

4. Quoted in *Niles Register, 10*:10, May 3, 1834.

5. *Correspondence of Andrew Jackson*, Edited by John S. Bassett, Carnegie Institute, Washington, D.C., 1931. See also the letter written in May of 1838 from George Strawbridge of Philadelphia to his son upon the news he was not on the *Oronoco* when it exploded on April 21st: "Thanks to a merciful Providence you are safe from the fatal accident of the *Oronoco*, & our great anxiety relieved. Tho you mention slightly, it ought to produce permanent & sincere Gratitude in every one of our Minds—near 300 have been murdered in this & the Cincinnati Affair [the *Moselle* explosion]—Most of these losses are owing to misconduct or ignorance & the Surviving Officers (& some of the missing Passengers) ought to be hanged. It was so near your fixed time of return & Route, we were very wary, tho the lists of sufferers did not include your name. As neither Congress nor the States will act any traveler must go in slow lines & wait for experienced old-fashioned Masters." [http://users.erols.com/aswhite/STEAMBOT.html].

6. Read on 12/3/1833, *Journal of the Senate*, 23rd Congress, 1st Session, pp. 6-19.

7. Peter J. Parish, Daniel Webster, New England, and the West, *Journal of American History, 54*, December 1967.

8. Robert V. Remini, *Daniel Webster, The Man and His Time*, W. W. Norton, New York., 1997.

9. Robert V. Remini, *Andrew Jackson and the Course of American Democracy, 1833-1845*, Harper & Row, New York, 1984.

10. Irving H., *Daniel Webster*, Norton, Bartlett, New York, 1978.

11. Martin Van Buren, *The Autobiography of Martin Van Buren*, John C. Fitzpatrick, Editor, Augustus M. Kelley, New York, 1969.

12. *The Papers of Daniel Webster, Legal Papers, Volume 3: The Federal Practice, Part II*, Andrew J. King, Editor, University Press of New England, Hanover, New Hampshire, 1989.

13. The General Improvement of Society, *Quarterly Christian Spectator, 6*, December 1834.

14. G. Edward White, *The Marshall Court and Cultural Change, 1815–1835*, Oxford University Press, New York, 1991. It was also this decision that caused the owners of the *Aetna* to transfer her from service in Philadelphia to service in New York where she exploded and initiated Vinton's first legislative action.

15. John K. Brown, *Limbs on the Levee: Steamboat Explosions and the Origins of Federal Public Welfare Regulation, 1817–1852*, International Steamboat Society, Middlebourne, West Virginia, 1989.

16. John McLure to Webster (1/14/1834) *Daniel Webster, Correspondence, Volume 3, 1830–1834*, Charles M. Wiltse, Editor, Published for Dartmouth College, Hanover, New Hampshire, 1977. The interesting thing to note in regards to collected letters and correspondence is that there are no letters about steamboat safety legislation in any of the collected letters of Grundy, Benton, or any other member of the 1837–8 Select Committee.

17. *Letter to the Hon. Daniel Webster on the Causes of the Destruction of the Steamer Lexington*, Charles C. Little and James Brown, Boston, 1840.

18. Daniel Webster: Economic Nationalism Continued, in *Jacksonian America, 1815–1840, New Society, Changing Politics*, Frank Otto Gattell and John M. McFaul, Editors, Prentice-Hall, Inc., Englewood Cliffs, New Jersey.

19. Quoted in *Niles Register, 1*:4, December 23, 1833.

20. Samuel Southard was the Chair, and members included George Bibb, Asher Robbins, Nathaniel Tallmadge, and Ezekiel Chambers.

21. John C. Burke, Bursting Boilers and the Federal Power, *Technology and Culture, 7*:1, 1966.

22. At least not until the final revised steam boiler inspection bill of 1852. To see an example of an Engineer's Certificate as required in the 1852 bill, see Douglas Stein, *American Maritime Documents: 1776–1860*, Mystic Seaport Museum, Mystic, Connecticut, 1992 (Figure 31).

23. Merrill D. Peterson, *The Great Triumvirate: Webster, Clay, and Calhoun*, Oxford University Press, New York, 1987.

24. William Nisbet Chambers, *Old Bullion Benton: Senator for the New West*, Little Brown, Boston, 1956.

25. Vinton on the floor of the House of Representatives, 5/18/24 as quoted in *Niles Register, 2*:11, May 22, 1824.

26. *Congressional Globe, 1*:4, December 23, 1833.

27. David John Denault, *An Economic Analysis of Steam Boiler Explosions in the Nineteenth-Century United States*, Ph.D. Dissertation, University of Connecticut, 1993. (Available from UMI Dissertation Services, Ann Arbor, Michigan).

28. David Lear Buckman, *Old Steamboat Days on the Hudson River*, Grafton Press, New York, 1909. What was true about the relationship between wrecks, the public hue and cry, and the resultant attempt at national legislation in 1824, 1830, 1833, and 1838 also proved true when an effective bill was finally passed in 1852: "these two accidents [the loss in 1852 of the *Henry Clay* by fire and of the *Reindeer* by boiler explosion], following so closely one after the other [July 28, September 4], resulted in a public agitation that secured the enactment of the Steamboat Inspection Bill of that year."

29. Report No. 478, Steamboats, May 18, 1832, in *Reports of Committees of the House of Representatives at the First Session of the Twenty-Second Congress, Begun and Held at the City of Washington, December 7, 1831*, Duff Green, Washington, 1831.

30. David F. Hawke, *Nuts and Bolts of the Past: A History of American Technology, 1776–1860*, Harper & Row, New York, 1988.

31. Edmund Burke, *On the Subject of Steam Boiler Explosions*, Executive Report No. 18, 30th Congress, 2nd Session, 1848 Serial Document 529.

32. Louis C. Hunter, *Steamboats on the Western Rivers: An Economic and Technological History*, Harvard University Press, Cambridge, 1949.
33. The Rev. James T. Austin in 1839 quoted in [34].
34. Hugo A. Meier, Technology and Democracy, in *Technology and Change*, John G. Burke and Marshall C. Eakin, Editors, Boyd and Fraser, San Francisco, 1979.
35. J. Aitken Meigs, speech in Philadelphia, July 13, 1854 quoted in [36].
36. Perry Miller, The Responsibility of Mind in a Civilization of Machines, in *Changing Attitudes Toward American Technology*, Thomas Parke Hughes, Editor, Harper & Row, New York, 1975.
37. Page Smith, *A People's History of the Ante-bellum Years, The Nation Comes of Age, Vol 4*, McGraw-Hill, New York, 1981.
38. Oliver Wendell Holmes, The Steamboat, in *The Knickerbocker, 13*, March 1839.
39. John F. Kasson, *Civilizing the Machine: Technology and Republican Values in America, 1776-1900*, Penguin Books, New York, 1976.
40. *The Mariner's Chronicle: Containing Narratives of the Most Remarkable Disasters at Sea*, Durrie and Peck, New Haven, 1834.
41. Charles Ellms, *Shipwrecks and Disasters at Sea*, S. N. Dickson, Boston, 1836.
42. S. A. Howland, *Steamboat Disasters and Railroad Accidents in the United States*, Dorr, Howland & Co., Worcester, 1846.
43. James T. Lloyd, *Lloyd's Steamboat Directory and Disasters on the Western Waters*, James Lloyd, Cincinnati, Ohio, 1856. (Facsimile reprint by Young & Klein, Inc., Cincinnati, Ohio, 1979.)
44. Helen Gordon, Tragedies at Sea and Disaster Averted in Victorian Parlor Songs, in *Manuscripts Collection*, Mystic Seaport Museum, Mystic, Connecticut, 1982. Usually, as Ms Gordon writes of the songs in the era under study: ". . . a shipwreck was usually a setting for some strengthening moral lesson or a chance for the Victorian listener to reaffirm his code of values and sense of Christian purpose," p. 23. Such a focus upon disaster in the words of a song is not that far distant from our own day when one recalls the Top Ten hit by Canadian Gordon Lightfoot in the 1970s that sang about the sinking of the *Edmund Fitzgerald* on the Great Lakes.
45. Katie Letcher Lyle's *Scaled to Death by the Steam: Authentic Stories of Railroad Disasters and the Ballads that were Written About Them*, Algonquin Books of Chapel Hill, Chapel Hill, North Carolina, 1991.
46. Related also to the sinking of the *Atlantic*, Geo. Dana Russell & Co., Boston, 1873.
47. *Commemorating the Terrible Disaster of Friday Night, September 7th, 1860*, Words and Music by Henry C. Work, J. L. Peters, New York, 1861. Lester S. Levy Collection of Sheet Music, Special Collections, Milton S. Eisenhower, Johns Hopkins University, Box 182, Item 30.
48. *Stanzas on the Occasion of the Destruction of the Forfarshire Steamship, September 7, 1838*, Davis & Horn, New York, 1839. Lester S. Levy Collection of Sheet Music, Special Collections, Milton S. Eisenhower, Johns Hopkins University, Box 56, Item 30.
49. Composed by W. F. Brough, Faulds, Louisville, Kentucky, 1856. Lester S. Levy Collection of Sheet Music, Special Collections, Milton S. Eisenhower, Johns Hopkins University, Box 182, Item 33.
50. Mary Wheeler, *Steamboatin' Days: Folk Songs of the River Packet Era*, Books for Libraries, Freeport, New York, 1969.

51. C. R. Roseberry, *Steamboats and Steamboat Men*, G. P. Putnam's Sons, New York, 1966. An interesting thing to note is the "inventor" of these barges was William C. Redfield who frequently appeared in Congressional collections of reports such as Report 478. Of course, as inventor of the barges and as founder and director of the Steam Navigation Company in the late 1820s and early 1830s, his frequent emphasis on steamboiler accidents can be seen as somewhat self-serving.

52. Carl D. Lane, *American Paddle Steamboats*, Coward-McCann, Inc., New York, 1943. (An especially good plate of an 1850 version of a safety barge, the *William H. Morton*.)

53. Fred Erving Dayton and John Wolcott Adams, (illustrator), *Steamboat Days*, Fredrick A. Stokes Co., New York, 1928.

54. Daniel Boorstin, *The Americans, The National Experience*, Vintage Books, New York, 1965.

55. W. H. Milburn, *The Lance, Cross and Canoe: The Flatboat, Rifle and Plough in the Valley of the Mississippi; The Backwoods Hunter and Settler, the Flatboatman, the Saddlebags Parson, The Stump Orator and Lawyer, as the Pioneers of Its Civilization; Its Great Leaders, Wit and Humor, Remarkable Extent and Wealth of Resources, Its Past Achievements and Glorious Future*, N. D. Thompson Publishing Company, New York, 1892.

56. *The Nature of Jacksonian America*, Douglas T. Miller, Editor, John Wiley & Sons, Inc., New York, 1972.

57. Report No. 478, Steamboats, May 18, 1832, in *Reports of Committees of the House of Representatives at the First Session of the Twenty-Second Congress, Begun and Held at the City of Washington, December 7, 1831*, Duff Green, Washington, 1831.

58. *Society, Manners and Politics in the United States: Being A Series of Letters on North America*, Weeks, Jordan and Company, Boston, 1839. A similar observation appeared in Mrs. Houston's *Hesperos, or Travels in the West* (1850) in which she commented about railroads: "I really think there must be some natural affinity between Yankee 'keep-moving' nature and a locomotive engine."

59. *Journal of the Franklin Institute, 10*, July 1830.

60. Bruce Sinclair, *Early Research at the Franklin Institute: The Investigation into the Causes of Steam Boiler Explosions, 1830-1837*, The Franklin Institute, Philadelphia, Pennsylvania, 1966.

61. Available today in *Franklin Institute and the Making of Industrial America* edited by Stephannie Morris (Microfiche Collection): Committee on Boiler Explosions, 1987.

62. Paul V. Anderson, *Technical Writing: A Reader-centered Approach*, Harcourt Brace College Publishers, Fort Worth, Texas, 1995.

63. On at least one occasion, the Secretary of the Treasury notes that a Franklin Institute report will be passed on to all members of Congress (Treasury Secretary Woodbury 12/17/36 letter to Franklin Institute on reception of General Report). [61]

64. Walter R. Johnson to the Hon. Levi Woodbury, Philadelphia, June 5, 1835. *Miscellaneous Letters Received*, Treasury Department. [61]

65. Hugh Richard Slotten, *Patronage, Practice, and the Culture of American Science: Alexander Dallas Bache and the U. S. Coast Survey*, Cambridge University Press, Cambridge, England, 1994: 66—"In addition to offering a means for advancing the 'cause of American science,' the position paid just as well as, if not better than, any

other scientific post in the country . . . equal to the salary of the principal government cabinet officers."

66. George E. Pettengill, Walter Rogers Johnson, *Journal of the Franklin Institute, 250*:2, August 1950.
67. Secretary of the Treasury, *Information Collected on Accidents on Board Steam Boats,* 1831: 21st Congress, 2nd Sess., Doc. 131, Serial Set 209.
68. Proceedings relating to the Explosion of Steam Boilers, *Journal of the Franklin Institute, 6*:1, July 1830.
69. Thomas S. Kuhn, *The Structure of Scientific Revolutions* (2nd Edition), University of Chicago Press, Chicago, 1970.
70. *Communications Received by the Committee of the Franklin Institute on the Explosions of Steam Boilers* in *Franklin Institute and the Making of Industrial America* edited by Stephannie Morris (Microfiche Collection) 1987: Committee on Boiler Explosions Fiche #148-9.
71. *Report of the Committee of the Franklin Institute of the State of Pennsylvania for the Promotion of the Mechanic Arts, on the Explosions of Steam-Boilers, Part I, Containing the First Report of Experiments Made by the Committee for the Treasury Department of the U. States* (1836). Available today in *Franklin Institute and the Making of Industrial America,* edited by Stephannie Morris (Microfiche Collection) 1987: Committee on Boiler Explosions Fiche #149–51.
72. Letter from Members of the Steam-boiler Committee to the Secretary of the Treasury, 9/28/1830 in *Franklin Institute and the Making of Industrial America,* edited by Stephannie Morris (Microfiche Collection) 1987: Committee on Boiler Explosions Fiche #149-51.
73. R. John Brockmann, Oliver Evans and His Antebellum Wrestling with Rhetorical Arrangement, in *Three Keys to the Past: The History of Technical Communication,* T. Kynell and M. G. Moran, Editors, Ablex Publishing Co., Stamford, Connecticut, 1999.
74. *Report of the Committee on Commerce, Accompanied by a Bill for Regulating of Steam Boats, and for the Security of Passengers Therein,* 18th Congress, Session 1, House Report No. 125, Gales & Seaton, Washington, 1824 (Serial Set 106).
75. Derived from *The Congressional Globe, 1*:4, p. 1, December 23, 1833.
76. Derived from *Journal of the Franklin Institute, 14*:4, pp. 217-222, October 1834.
77. Derived from original proposed Bill as printed in *The Congressional Globe, 6*:1, December 11, 1837.
78. Charles F. Penniman, The First Materials Testing in America, *ASTM Standardization News,* January 1991.
79. William H. Hale, Letter To The Hon. Felix Grundy of the U. S. Senate, May 13, 1838, *Journal of the American Institute* in September 1838, Vol. III, No. 10, pp. 529-534. Reply to T. B. Wakeman's critique of the 5/13/1838 letter, *Journal of the American Institute* in September 1838, Vol. III, No. 12, pp. 651–656.
80. John D. Ward, T. B. Stillman, and Stephen Dod, Report on the Causes, &c, of the Explosion of Steam-Boilers, *Journal of the American Institute* in September 1838, Vol. III, No. 12. The American Institute in New York City, a sister institution to the Franklin Institute, also began investigating steam-boiler explosions in July 1837 and published their findings in September 1838.

81. Steamboats on the Western Waters, *Journal of the Franklin Institute, 14*:5, November 1834. Locke made his investigation into the explosion and wrote his report immediately after the sinking of the Moselle in April of 1838. He delivered it to the Mayor of Cincinnati in June, and the mayor was taken aback by Locke's indictment of the rich merchants of Cincinnati pushing for "fast" rather than "safe" running of steamboats. "To those acquainted with the early mercantile history of our country, when it was no uncommon thing for a party of merchants to be detained in Pittsburgh from six weeks to two months by low water and ice, the existing state of things is truly gratifying. The old price of carriage of goods from the Atlantic seaboard to Pittsburgh was long estimated at from five to eight dollars for one hundred pounds. We have instances in the last five years of merchandise being delivered at the wharf of Cincinnati, from Philadelphia by way of New Orleans for one dollar per hundred." As a result, the City Council pigeon-holed the report until Locke's protests and those of the citizens of Cincinnati forced the City Council to release the report in the Fall of 1838.

82. Fredrick Way, *She Takes the Horns: Steamboat Racing on the Western Waters,* Cincinnati, 1953.

83. *Report of the Committee Appointed by the Citizens of Cincinnati, April 26, 1838, to Inquire into the Causes of the Explosion of the Moselle, and to Suggest such Preventive Measures as May Best Be Calculated to Guard Hereafter Against Such Occurrences,* Alexander Flash, Cincinnati, Ohio, 1838.

84. *Steam Navigation Part I, The Boiler,* Ridgeway, London, 1832.

85. Evidently Renwick and Bache personally knew each other for some time as Renwick wrote in his 10/23/1843 letter of support for Bache's application to be Superintendent of the Coast Survey. (Smithsonian Institution Archives, Record Unit 7053, Bache Papers, Box 6).

86. James Renwick, *Treatise on the Steam Engine,* Carvill & Co., New York, 1839.

87. John Harmon Ward, *Steam for the Million: An Elementary Outline Treatise on the Nature and Management of Steam and the Principles and Arrangement of the Engine Adapted for Popular Instruction,* [earlier entitled in 1845: *An Elementary Course of Instruction on Ordnance and Gunnery: Prepared for the Use of Midshipman at the Naval School, Philadelphia: Together with a Concise Treatise on Steam Adapted to the Use of Those Engaged in Steam Navigation*] Carey and Hart, Philadelphia, 1846, 1847, 1860, and after Ward's death in 1862, 1864 and 1867.

88. Jacob Walter, *Essay on Steam,* Document No. 173, 25th Congress, 3rd Session, Serial Document 347.

89. *Relative to Steamboat Explosions,* Document No. 145, 27th Congress, 3rd Session, Serial Document 429.

The Gold Dust Fire

Chapter 37. The End of the "Gold Dust"

For, three months later, August 8, while I was writing one of these foregoing chapters, the New York papers brought this telegram:—

A Terrible Disaster

Seventeen persons Killed by an Explosion on the Steamer "Gold Dust."

Nashville, Aug. 7—A Dispatch from Hickman, Ky., says:—

"The steamer 'Gold Dust' exploded her boilers at three o'clock to-day, just after leaving Hickman. Forty-seven persons were scalded to death and seventeen are missing. The boat landed in the eddy just above the town, and through the exertions of the citizens the cabin passengers, officers, and part of the crew and deck passengers were taken ashore and removed to the hotels and residences. Twenty-four of the injured were lying in Holcomb's dry-goods store at one time, where they received every attention before being removed to more comfortable places."

"Gold Dust Fire" song [1, p. 41].

A list of names followed, whereby it appeared that of the seventeen dead, one was the barkeeper; and among the forty-seven wounded were the captain, second mate, and second and third clerks; also Mr. Lem. S. Gray, pilot, and several members of the crew.

Mark Twain, *Life on the Mississippi* [2]

Endnotes

1. There is also an African-American spiritual song that recounts this disaster and is in Mary Wheeler, *Steamboatin' Days: Folk Songs of the River Packet Era*, Books for Libraries, Freeport, New York, 1969.
2. Later while working as a reporter for the *San Francisco Daily Morning Call*, Clemens wrote an article on September 7, 1864 about the explosion of the Washoe's boilers which killed one hundred passengers.

CHAPTER 4

Steamboat Politics and Rhetoric

The public feeling demands some legislation on the subject and it must be given.

George Curtis to N. R. Knight letter,
Papers of the Select Committee on
Bill S. 1, 25th Congress, Session 2 [1]

"Steamboat Explosion" by J. N. Bang
(on the occasion of the explosion of the *Moselle* at Cincinnati,
April 25th 1838)

An attempt has been made in the following lines to portray the scene in such vivid colors, as to produce a deep impression on the minds of all who may read them; and help to discountenance, it is hope, and, if possible, put a stop to such reckless conduct and inexplicable faulty order on the part of those, who have the control and management of Steamboats.

Dismembered bodies strewed the ground
Heads arms and legs were scattered 'round,
A melancholy sight!
From dead and dying streamed the blood,
Which tinged the waters of the flood,
And almost raised their height.

The steamboat sank beneath the wave,
And all on board there found a grave,
Within the watery deep:
Appalling yells and deafening cries
From those engulfed to heaven arise,
Till hushed in final sleep

This sad disaster loudly calls
On those convened in Congress halls,
Their power to interpose;
By strong enactments to restrain
All rash "experiments" again,
Which have a fatal close.

May 11, 1837, Thirty Miles South of Natchez

In the Washington, D.C. newspaper of record, *The National Intelligencier,* the following was published:

> One of those terrible accidents but too common on the Western rivers occurred on Tuesday at one o'clock in the morning, by which one hundred and fifty lives were lost. The steamer *Ben Sherrod*, Captain Castleman, left this place [New Orleans] on Sunday morning bound to Louisville, and at the time just mentioned, when about 30 miles below Natchez, she was found to be enveloped in flames, and out of near 200 persons on board, only about 50 or 60 were saved [2].

Later reports on the *Ben Sherrod* (see Figure 24) appeared in three other issues of the Washington paper, and it became apparent that the *Ben Sherrod* had been racing, had overheated her boilers, and had an intoxicated crew. As commentary on this particular accident, the following appeared in the March 18th edition of the *Niles National Register*:

> The melancholy event is termed in the Vicksburg paper an "accident." That little word speaks the public feeling in relation to such wanton sacrifices of human life, more truly than the most elaborate discussion could do and must convince all person that public opinion, or the voice of philanthropy, cannot correct the murderous practices of rival steamboat commanders and owners, to whom the lives of passengers are nothing, compared to the mighty object of arriving five minutes before an opposition boat. We are not advocates of

Figure 24. The *Ben Sherrod* blows up.

"Lynch law" in any shape, but think, if it can be justified, it ought to be executed on such captain-murders [3, p. 34].

A Brief Coincidence of Political Interests

Only months after his inauguration, beset by a financial panic and with deep divisions within his own Democrat Party, Martin Van Buren suffered in the Fall midterm elections in 1837. The Whigs had even realized a stunning victory in Van Buren's own home state of New York, a fact that Van Buren called the "New York tornado" [4, p. 99]. In addition to suffering in New York, Van Buren's Democrats had also lost electoral ground in the West, an area which contained one-fifth of the total U.S. population [5, p. 137]. Consequently, in November, John C. Calhoun of South Carolina, Van Buren's nemesis for years, was publicly writing about Van Buren's weakened political condition:

It is clear that with our joint forces [Whigs and nullifiers] we could utterly overthrow and demolish them [6, p. 247].

Within this besieged setting, Van Buren set about drafting his December State of the Union Message. Moreover, it was not surprising that very much like his mentor Jackson's message in 1833, Van Buren offered the following as his penultimate paragraph in his nineteen page message:

The distressing casualties in steamboats, which have so frequently happened during the year, seem to evince the necessity of attempting to prevent them, by means of severe provisions connected with their customhouse papers. This subject was submitted to the attention of Congress by the Secretary of the Treasury in his last annual report, and will be again noticed at the present session, with additional details. It will doubtless receive that early and careful consideration which its pressing importance appears to require [7, p. 19].

This fourth attempt at legislation for steamboats took a decidedly different turn within 24 hours. Rather than waiting weeks, months, and even years for a committee to collect information and produce a bill, Felix Grundy quickly proposed specific steamboat legislation (S. 1 *to provide for the better security of the lives of passengers on board of vessels propelled, in whole or in part, by steam*) ready for consideration and amendments. Moreover, with his decades of legislative experience, Grundy decided to take no chances with his bill in any committee currently constituted even though his and the president's party controlled the Senate 30 to 22. Instead, Grundy asked for the creation of a Select Committee such as he had used earlier in February 1834 to consider the "compromise tariff" that broke the Jackson/Calhoun/Webster nullification impasse. Moreover, Grundy's earlier Senate committee had even included a number of the same senators he recruited now: the Whig leader Webster, the Democrat Calhoun, and himself, the Democrat stalwart, as Chair [8, p. 384; 9, p. 39]. He also added a number of other senators: Senator Benton of Missouri, a western Democrat stalwart like himself, and a Democrat to represent the north, Senator Garrett Wall of New Jersey, as well as a Democrat to represent the Deep South, Senator Robert Walker of Mississippi. Grundy also included the relatively new senator from Delaware, Thomas Clayton, perhaps hoping Thomas might prove as effective as his cousin, John, had proved in Grundy's earlier Select Committee.

The Select Committee

A hundred and fifty years later, one can get a sense of Grundy's Select Committee from a little book published in 1839 by the anonymous, "A Looker on Here in Verona," entitled *Sketches of United States Senators, of the Session of 1837–8.*

Of Mr. Webster, the Looker described him as:

> . . . short, large, heavy and unwieldy; in movement he is slow and apparently inactive. He dresses plainly, in dark colors, with neatness and taste, but without any attempt whatever at display. His countenance is very remarkable, his complexion saturnine, his eyes and hair of a deep black. His lips are thin, his teeth of dazzling whiteness. His forehead is very peculiar, of most uncommon magnitude, his brows heavy and lowering. The moment that your eyes rest upon him, you conclude that he is a man of great mind, and conscious of intellectual superiority [10, p. 18].

Webster was very interested in seeing steamboat legislation passed by Congress as could easily have been inferred from his 1833 senate speech, and by his longtime involvement in steamboat law since 1824 when he had argued the landmark Gibbons vs. Ogden case in the Supreme Court. Webster also believed that there was legal precedent for such a law as he alluded to in his speech:

> And in the first place, I think the boat itself should be made subject to forfeiture, whenever lives were lost through the negligence of those conducting it. There is nothing unreasonable in this; analogous provisions exist in other cases. The master of a merchant ship, for instance, may forfeit the ship by a violation of the law, however innocent the owners may be, even though that law be only a common regulation of trade and customs [11, pp. 765-766; 12, p. 647].

With another run for the Whig presidential nomination coming up, and having completed a recent swing through the West which was so dependent on steamboat travel and which held so many new voters, it was probably most opportune for Webster to support a steamboat bill . . . even if the new bill had been proposed by the opposing Democrat party.

Yet, of the 13 parts of Democrat Grundy's proposed bill, nearly half, six, were identical to Webster's 1833 bill outline. Especially in the civil liability and licensing area, the two draft bills seemed to be in close agreement. Moreover, Grundy needed Webster's help in this matter because he knew that if the Whigs combined with Calhoun's nuillifer Democrats, the Administration would not be able to pass a bill in the Senate.

Webster had locked horns in previous Congresses with his fellow committee member Calhoun over nullification and with his fellow committee member Benton in the compromise tariff and in the "expunging" bill. Yet, Webster had also worked with them on occasion. He had, for example, joined Calhoun when Calhoun had denounced the dictatorial tendencies of President Jackson [9, pp. 157-158]. However, Webster and Benton had a quite present disagreement in that 25th Congress for there was a personal financial interest in this bill by this eastern senator.

At a time when it was considered quite proper for politicians to profit from relationships to bankers, industrialists, and real estate syndicates, Daniel Webster and his fellow senator from New Jersey and chair of the 1834 Naval Affairs Committee which had last proposed a steamboat bill, Samuel Southard, were members of a real estate syndicate. Their syndicate had purchased rights to a 536,000 acre tract of land in Missouri and Arkansas, the Clamorgan Grant, and thus, both senators would personally benefit from easy and safe transportation of commerce and settlers to their grant via steamboats. However, when Southard and Webster tried to clarify rights to their grant in Senate Bill 153 during this same session of Congress, they crossed swords with Committee members from the West, Benton of Missouri and Robert Walker of Mississippi. These two senators from the West were both actively supporting initiatives to make such western lands free and open to squatter settlers (pre-emption legislation) [13, p. 242]. Consequently, even though Webster was surely interested to see a bill issue from the Select Committee, and even though he brought to bear his power and influence, it was all accompanied by a lot of political baggage.

Of Mr. John C. Calhoun, the Looker in the *Sketches* described him in the following way:

> While addressing the Senate he stands quite erect. His eyes are fixed upon the carpet. His usual action is that of the right hand up and down . . . His figure is tall, his hair bushy and abundant, his forehead not remarkably high, but broad and compact. His countenance altogether has a Roman cast and expression. His lips are stern, boldly outlined and generally closely compressed; his eyes dark and keen. In argument or debate he never refers to a note; what he says he says right on—his ideas appearing to crowd more rapidly than language can be found to give utterance to them. Although wanting the graces of manner, no man is heard with more attention [10, p. 25].

As with all the others, his alliances and allies had moved him in and out of confederation with others on the Committee. Losing the Vice Presidency to Van Buren with Van Buren's follow-up election as President was a bitter blow to Calhoun's pride [14, p. 219]. But Calhoun realized that his nullifiers held the balance of power in the Congress and that more would be gained by selectively allying with the administration on various bills. For example, he had supported the administration in their earlier banking bills, and, in turn, the administration had supported him in his anti-abolitionist petitions in December and January.

Moreover, just when Van Buren and Grundy initiated steamboat legislation because of national concerns in the 25th Congress, steamboat legislation took on a decidedly local South Carolinian concern for Calhoun.

In October 1837, the Southern Steam Packet Company's *Home* sank (see Figure 25) on its return trip to Charleston, South Carolina, causing 95 people to drown off Cape Hatteras and undermining public confidence in East coast scheduled steam service [15, pp. 13-37; 16, pp. 439-440]. The sinking grabbed the attention of Charleston (see Figure 26) resulting in such things as a sermon printed just like one had been 13 years earlier in New York City with the *Aetna* explosion. *The Voice of God In Calamity* had this as its purpose:

> As these proceedings are to all be made public, and will, doubtless, occupy much of the public attention, it may not be unadvisable to consider the matter religiously, and to hold up to the general view those "lessons of eternity" which are, surely, no less necessary to save us from "making shipwreck" of our future and everlasting hopes. While we thus hear the voice of God and the voice of man teaching and admonishing us; while we are thus led to humble ourselves under the mighty hand of God, and to protect ourselves from the inexcusable perils to which we are exposed by the cupidity, or experimenting boldness, or the reckless indifference, of men, we may hope that this whirlwind calamity, however desolating in its progress, and heart-rending in its consequent misery, will leave behind it, an atmosphere purified, a sky cloudless, and a city rejoicing in hope of future safety [17, pp. 3-4].

The local public pressure on the Charleston city mayor, Henry Pickney, motivated him to create Charleston's own municipal select committee, the Committee of Twenty-one (much like a committee twenty years earlier in Philadelphia), and to pass a law regulating steamboats. This Charleston law required steamboat inspections in November 1837—a requirement that few other cities had [18]. Not only did Charleston pass the law, but they produced a report and sample legislation and circulated their material to members of their Congressional delegation—including Calhoun [19; 20, pp. 3-7].

It is interesting to consider the elements of the Charleston proposed regulations that included [19, pp. 35-36]:

> First. Whether it be not practicable by legislation, or some other means to compel greater strength in the construction of steamers, and increased comfort and security to passengers.
>
> 2ndly. To look to the proprietors of the steamers for redress in every case of disaster occurring from the weakness, or unseaworthiness of the vessel, the ignorance, incompetency, or immoral habits of the Captain or the Engineer. That they should be restricted in the number of passengers, on each trip, taking no more than they can safely and conveniently accommodate with berths, or

Figure 25. The sinking of the *Home*.

Figure 26. "I Love Thee Dearest Brother"—Music Occasioned by the Sinking of the *Home*: "These words were written by Miss Cynthia H. Stow who was lost in the wreck of the *Home* in 1837 and left for her brother a few hours before she embarked under the impression that she would never meet him again."

sleeping places. That passenger boats should carry no freight, or where they carry freight, fewer passengers. That each Steamer, should be compelled to have ready for immediate and faithful service, a number of strong and large boats, sufficient to remove and rescue passengers and crew. That they be provisioned with a double allowance of food and water on each trip. That wherever it can be clearly proved, that the Captain or Engineer of a Steamer, have caused her loss, or the destruction of human life, by incapacity, intemperance, or neglect, that he or they shall be made liable to criminal prosecution, and be held forever after as incapable of command or trust in a similar capacity.

Lastly. Your Committee would recommend, That a strong appeal be made through the MEDIUM OF THE PRESS, by the EDITORS in this City, and in our sister Cities, to the moral sense of the Owners, Commanders, and Engineers of the Steam Packets now engaged in the Trade, earnestly and emphatically calling upon them as Christian men, as responsible and moral agents, voluntarily assuming the trust of human life and human happiness, to guard them with all care and diligence, and to save us from a repetition of those agonizing scenes, which have cast the deep shadows of despondency over our City . . .

Elsewhere in the Charleston Committee report were these two additional regulations

- the appointment of mechanics skilled in steam and boat machinery as well as experienced ship builders to minutely and particularly examine the machinery and other equipment and to certify to the master that she is or is not seaworthy [19, p. 5]; and
- that this testing and examination produce a certificate of the ship's seaworthiness [19, p. 5].

These local South Carolina laws most definitely made it to the eyes of Calhoun and probably the entire Senate Select Committee because one of the letters held in the National Archives file of the Senate Select Committee was written by Thomas Lee, a U.S. District Judge in South Carolina and a key member of the Charleston Committee of Twenty-one. Lee's letter to the Senate Select Committee and conveyed by Representative Hugh Legaré, one of the South Carolina House members, and began in the following way:

I observe that there is a Bill before Congress on the subject of Steam Boats intended to protect Passengers on such Vessels from the Calamities to which they are too often exposed by the want of the necessary Precautions for their Safety. I rejoice that this Business is before you and I have no doubt by proper Legislation that tho[u]sands of lives will be saved [21, p. 770].

In his letter, Judge Lee added a few recommendations to the ones listed in the published Committee of Twenty-one Report. Lee additionally suggested:

7. that lights be put on the bow and stern of vessels after sunset
8. that when collisions seem imminent that the vessels steer off one or two points so as to miss each other.

Thus, in Calhoun's political calculations, the local South Carolina pressure that occurred because of the sinking of the *Home* balanced the potential in the S.1. bill of expanding the power of the Federal government at the cost of state power—just exactly what Calhoun and the nullifiers were traditionally against. (The nullifiers believed that the individual states could suspend a federal law within their state boundaries since the federal government was only a league of freely associated states.) And, in fact, in the Charleston report from the Committee of Twenty-one distributed to the Congress [19, p. 25], there was an explicit discussion of this state versus Federal power aspect in the proposed regulation of steam navigation:

> The Law on this subject must be general. The separate action of any one State in the Union, would be nugatory, and inefficient. Nor is there great room to be hoped for, or expect much benefit from the action of the City Council of Charleston, unless the constituted authorities of our sister cities will co-operate, and adopt similar plans for examining and scrutinizing, and reporting on the condition of every boat in the service. The action must be universal, to be really beneficial. How far under the federal constitution, and how beneficially the national Legislature, can act in this matter, it is not for your Committee to say; but assuredly that honorable body must have power under some article of the Federal Constitution, to guard, and protect the lives of American citizens, from the willful and wanton destruction to which they are liable, under the present loose, and unregulated system of steam navigation [19, p. 27].

Later, during the House debate on the Steamboat Bill, one of the members of Calhoun's South Carolina Congressional delegation, Representative Campbell, discussed this state versus Federal power aspect:

> . . . He pressed the constitutional objection which had been urged when the bill was in [House] committee. The certain interposition of the Federal courts was not a sufficient argument in favor of a law in itself unconstitutional. The General Government could have no jurisdiction over steamboats moving within a State. The same doctrine would give the Government jurisdiction over every wagon on every road in every State of the Union, and the Government was already a consolidation [22].

There was also another ramification of this local South Carolina concern for steamer regulations that Calhoun needed to consider. The *Home* did not sink from a boiler explosion but rather from faulty construction, and thus most of the steam boiler technical and operational solutions present in all the previous four steamboat bills were not present in the Committee of Twenty-one November 1837 recommendations. However, there was some coincidence between the Committee of Twenty-one's and Judge Lee's proposal in general fire-fighting and safety solutions and a lot of coincidence between the civil liability and licensing solutions with those proposed previously. Thus, this Charleston report coming into Calhoun's office as the last outside report to reach him during the Senate Select Committee's considerations, and accompanied by a letter from Judge Lee of the South Carolina Committee, probably would better prepare Calhoun to support general fire-fighting and safety solutions and the civil liability and licensing solutions rather than the specific technical and operational solutions proposed by the Institute's reports [23].

Calhoun may also have had an important political reason to get involved in this Federal regulatory legislation which was so uncharacteristic of him as a nullifier. In 1837 and 1838 Calhoun led his fellow nullifiers into playing with both the Democratic Van Buren Administration and the Whigs as a wild card in their power struggles. Calhoun did this because he reasoned that he could ensure a key role for his Democrat conservative nullifiers. It had already worked in that he had supported the Administration in their banking legislation, and the Democrat Administration had supported him in his pro-slavery responses to the Vermont abolitionists [24]. Such political zigzags did mean that he would occasionally have rather odd political bedfellows like the northerners Van Buren and Webster and that Calhoun might even appear to compromise his conservative principles as he moved his influence back and forth. This became especially evident when he supported the Administration banking legislation earlier in the Fall of 1837.

Calhoun clearly understood that to maintain his power and influence in this zigzag game, he had to rely upon his South Carolina Congressional delegation to loyally support him through all these twists and turns, and many claimed that Calhoun was a virtual dictator of politics in South Carolina [25, p. 209; 26, p. 4]. Yet, in the Banking debate earlier in the Fall of 1837, he did not receive that kind of unswerving support from four in his delegation: Senator Preston, Representative Campbell, Representative Hugh Legaré—who had received the pro-steamboat letter quoted above—and Representative Waddy Thompson who represented Calhoun's own district. In a letter to his brother in April of 1838, Calhoun wrote regarding Thompson:

> In fact, if I mistake not, there is much discord in embryo in the North, political and otherwise. Now is our time. What a misfortune, that we are divided among ourselves. Preston and Thompson have done much mischief—more than they can repair if they live one hundred years [27].

Calhoun planned to purge Thompson and Legaré who were up for re-election in the 1838 Fall elections, and, in the very personal attacks by Calhoun on Thompson, only a public exchange of letters prevented a duel [25, pp. 244-247; 28; 29].

The House of Representatives had been considering steamboat legislation since December 1836, and Rep. Waddy Thompson (see Figure 27) from South

Figure 27. Representative Waddy Thompson [26, p. 1].

Carolina was a member of the **House** Select Committee considering steamboat legislation. Thus, in addition to all the usual claims that Thompson could plan to make in that 1838 Fall election about taking care of Carolinian concerns in Washington, what better platform to be handed by political fate than to be a member of the House Select Committee overseeing steamboat legislation just at the time when South Carolina became a beehive of activity in steamboat legislation prompted by the local *Home* sinking. Moreover, what better political platform for Calhoun to assume to defeat Representative Thompson than to become a member of the **Senate** Select Committee overseeing steamboat legislation. And, it just so happened that the person picking senators to be members of the Senate Select Committee was none other than Calhoun's very old friend, and the person who had married his second cousin, Senator Felix Grundy.

The *Sketches of United States Senators, of the Session of 1837–8* described Senator Grundy in the following way:

> Rejoices in a good portly figure, no ways discrediting the soil which may have given him sustenance, with a countenance rosy, round and smooth, expressive of good humor, good feeling and self-contentment—eyes gray and sparkling—hair whitened, but bushing about in every direction and abundant; neither toil nor thought has chiseled cheek or brow. His age, with the apparent healthfulness and comfort which accompany it, deserves the beautiful description of the finest of poets, "frosty but kindly" [10, p. 38].

Like Webster, Clayton, Calhoun, and Benton, Grundy had been a lawyer before being a politician. He had also been a loyal supporter of both the Jackson and Van Buren Administration, having been first appointed to the Senate in 1829 through Jackson's discrete efforts [30, p. 173]. Later, in 1838, Grundy would be appointed Attorney General by Van Buren and function as a campaign adviser in the 1840 presidential election.

Grundy (see Figure 28) had long been a close friend to Calhoun, having arrived in Washington D.C. with him some 20 years prior [31, p. 3]. Subsequently, Grundy had married a second cousin of Calhoun and had supported Calhoun continuing as Vice President when Jackson worked to move him out of his position in 1824-5 [30, pp. 174-175]. Thus, Grundy moved in the Democratic middle between the Calhoun conservative nullifiers and the Jacksonian Democrats. Such a medial role for Grundy was best demonstrated during the infamous 1830 Jefferson Day toasts when first President Jackson raised a toast for "The Federal Union: It must be preserved!" to which Calhoun retorted with the toast "The Union: Next to our Liberties, the most dear!" It fell to Grundy to finish with a toast attempting to reinstate party harmony by saying: "The Republican party throughout the nation: May they be as harmonious in action as they are united in principle!" [30, p. 176].

Grundy's loyalty to the administration was sufficient enough for him to be the designated point person in the S.1. Bill. However, he may have been designated leader on this bill dear to the hearts of his western Tennessee constituents dependent upon steam travel to palliate them for his less favored votes on banking bills earlier in the legislative term.

Of Mr. Benton, the Looker in *Sketches* described him as:

> . . . a personage of goodly corporeal dimensions—florid of complexion, and of marked pinquidity. . . . dress[ing] with particular neatness and care, to use in a

Figure 28. Senator Felix Grundy, Chair of the Senate Select Committee.

style of apparent dandyism the eyeglass which hands upon his bosom in fetters of gold to be almost perpetually engaged in writing or in the examination of the heaps of documents and papers piled upon his desk, a finely attired, pleasant-seeming gentleman. . . . Unmeasured vituperation has been poured upon him, but on the other hand his speeches are reprinted most extensively, and further and more widely than his enemies can, or will realize, is his name with the people a "household word" [10, p. 29].

Like Grundy, Benton had been counted upon by Jackson and Van Buren to be a loyal supporter of Administration bills. Benton was so loyal to his fellow Westerner and Democrat leader, Jackson, that upon Jackson's retirement, Benton led the fight in the Senate to expunge the Whig's earlier censure of Jackson's imperious gestures, the famous "expunging" bill: "solitary and alone I set this ball in motion" [10, p. 32]. Furthermore, Benton was so loyal to the succeeding Van

Buren administration that he was named by a Mississippi Democrat convention to be Van Buren's Vice President [32, p. 158].

Benton also stood together with Van Buren's archenemy, Calhoun, earlier in the Fall of 1837 on the banking bills. Public opinion in Missouri also supported steamboat legislation as can be sensed in this December letter from John Daggett of St. Louis and received by the Senate Select Committee:

> The importance of the subject is now sufficiently well understood by all classes, and especially by owners, shippers, and passengers, to render it unnecessary to say any thing on that head, the necessity of some laws or general regulations must be apparent to every one, they should be general & uniform throughout the United States, this uniformity could never be obtained by enactments of the several state Legislatures, even if it could, they would be the question of jurisdiction continually arising [21, p. 767].

Moreover, in this letter from a constituent in Benton's state, the emphasis—very much like that of the Charleston Committee of Twenty-one—was on the general fire-fighting and safety and on the civil liability and licensing solutions rather than on the specific technical and operational solutions.

As one of peripheral standing among the Senate Whigs, Mr. Thomas Clayton of the Senate Select Committee was not described in the *Sketches*, and, when he did surface in a book, the notice was rather small:

> [T]o John's [John Middleton Clayton] left was the "other Clayton," a very quiet, tobacco-stained man who simply "chews and votes, and votes and chews" [33, pp. I, 168, 177, 179].

In December 1836, the better known cousin, John Middleton, had for the moment, tired of national politics and switched places with his cousin who was Chief Justice of Delaware [34, p. 124]. This was the second time that Thomas had filled out an unexpired Senate term. The earlier one occurred in 1824 when Vinton had first raised the steamboat problems in the House.

When Judge Clayton appeared during the debate on S.1. after it reached the Senate floor, the only words of his recorded in the debates came when he noted that certain amendments were deemed "innovations on common law." He also successfully fought against amendments having to do with "racing" since that would be a very difficult thing to prove in a court of law. Thus he brought a lawyer's eye to the debate on S.1.

Of Select Senate Committee member Garret Wall, the Looker in *Sketches* described him as:

> In person he is compactly and rather heavily constructed. He usually wears a frock coat—gray, with standing collar, and of staid appearance. This gives him an aspect rather military. His countenance betokens him intelligence—his features not particularly marked—his eyes are blue and his hair abundant. . . . He reasons well; his efforts are enlivened and adorned with classical allusions,

and his embelishments, although frequent, and from various sources, are regulated by a sound and judicious taste [10, p. 43].

Wall was a little known New Jersey Democrat who generally supported the Administration. Just at the time of the Committee considerations, Wall had probably incurred the ire of Calhoun because Wall had presented a petition from 115 women in New Jersey in support of the move for the abolition of slavery [35, p. 328].

At 35, Committee member Robert Walker of Mississippi was the second youngest member of the Senate, but, as a representative from a Western state, he was a force to be reckoned with in all matters western [36, p. 55]. He pushed for recognition of Texas independence, and was especially keen on turning over public lands to the settlers, some called "squatters," who now lived on the lands. The "pre-emption" land bill was Walker's signature in the previous 24th Congress.

He had opposed sending the pre-emption bill to a hostile committee, so he maneuvered Van Buren into creating a Select Committee with him as chair to consider it [36, p. 68; 37; 38, p. 74]. The bill issuing from his Select Committee was being acted upon by the same Congress at the same time as S.1. steamboat bill. Walker's land bill was passed by the Senate on January 23rd 1838 with a Democrat to Whig partisan split [37, p. 26]. So Walker had some knowledge of Select Committees and was interested in all matters Western such as this steamboat bill. Moreover, if the Administration and their Senators could support his pre-emption bill, perhaps he could support their steamboat bill.

Walker found common cause on this pre-emption topic from the Massachusetts Whig leader, Daniel Webster. Webster said about himself "I am as Western a man, on this [Western internal improvements bill] as he among them who is most Western." Webster came out most strongly for pre-emption in the years 1837 to 1840 [39, pp. 537, 541], as did Grundy who also voted to support Walker's bill [31, p. 86]. As odd bedfellows as Walker and Webster might have been, one would have expected Walker and the Missourian Benton to get along. However, in the Spring of 1837 in the debate regarding the pre-emption bill, Benton had cried out on the Senate floor "God save the country from the Committee on Public Lands" [the Committee Walker chaired] to which Walker retorted "God save us from the wild, visionary, ruinous, and impracticable schemes of the Missouri Senator" [36, pp. 73-74]. Relations between the two were so bad that several southern newspapers published rumors of a "fatal rencontre" between the two [36, p. 74].

The Initial Proposed Bill in December 1837

Into this potent mix of divergent Committee political principles, pressures, power, and goals, Grundy presented a bill that can be compared quite easily with the three previous bills as shown in Table 4. In essence, Grundy's proposed bill offered no new items, but rather a collage of what had been proposed in the three preceding attempts at legislation. As with all the others, Grundy's bill called for periodic inspections, certificates of boiler pressure limits, and licenses. As in Wickliffe's bill and the Naval Affairs Committee's bill, as well as in the Charleston South Carolina report, the Grundy bill called for various operational and general fire-fighting and safety solutions. As in Webster's proposal, Grundy's bill had various civil and liability penalties. In fact, the astonishing speed with which Felix Grundy was able to place a proposed bill before the Senate—within 24 hours after Van Buren's State of the Union message calling for it—was perhaps less astonishing when one noted that Grundy's bill was nearly identical to the Senate Naval Affairs Committee's three years prior.

Moreover, like Wickliffe's and the Naval Affairs Commitee bill, the initial Grundy bill included two operational rules—including lines 7 to 12 in Section 7 that called for the continued activation of the water pumps even when a steamboat was stopped (see Figure 1). One other asset that Grundy's Committee had which no other Congressional Committee had ever had was full, complete, and comprehensive technical information on steamboiler explosions in the form of the series of reports submitted by the Franklin Institute.

THE BILL REPORTED OUT OF COMMITTEE

May 8, 1838
Dear George
... The three steamboat disasters, excite a strong feeling! Will it do any good, or procure any Law [45].

When Grundy presented S.1. with its amendments on the floor of the Senate on January 8, 1838, the Bill was drastically changed from originally proposed one modeled on the Committee on Naval Affairs Committee's prior bill (see Table 5).

In essence, the technical and operational solutions were largely dropped and the general safety and fire-fighting and civil liabilities remained. Why? One reason may be that most of the outside letters submitted to the Senate Select Committee were critical of the technical and operational solutions. Grundy himself, in the opening of the considerations of the bill on the floor of the Senate, said:

The extensive and minute measures which had been taken by the committee to obtain information on the subject; a most important portion of which, from masters and engineers of steamboats, had arrived since the bill was framed by the committee. It was by this information that the amendments proposed had been suggested [47, p. 3].

Table 4. Comparing Grundy's 1837 Bill to the Previous Three Bills

	1824 Vinton's Proposed Law [40]	1830-2 Wickliffe's Proposed Law [41]	1833-4 Webster's Proposed Law [42, p. 1]	1834 Committee on Naval Affairs Proposed Law [43, pp. 217-222]	1837 Grundy's Proposed Law [44, p. 8; 24, Appendix B]
Technical Solutions					
Boiler and machinery inspections	X	X	X	X	X
Certificate of boiler pressure limits	X	X	X	X	X
Two safety-valves in place	X	—	—	—	—
One of two safety-valves locked	X	—	X	—	—
Pressure gauge included	—	—	X	—	—
Inspectors shall test boiler pressures	X	—	X	X	X
Periodic testing shall take place	—	X	—	X	X
Operational Solutions					
When boat is stopped, machinery will be kept in gear so that water pump is kept activated, and the safety valve must be kept open	—	X	—	X	X
Must shut off steam when passing ascending vessels at night	—	X	—	X	X

General Fire-fighting and Safety Solutions

Solution				
Must carry long boats for safety	—	—	X	X
Must carry suction hose and fire hose	—	—	X	X
Night running requires signal lights	—	—	X	X

Civil Liability and Licensing Solutions

Solution				
Licensing of all steamboats	X	X	X	X
Transportation of gunpowder aboard ship requires special placement and handling	—	—	X (a singular item in reaction most probably to Lioness disaster)	—
Tampering with safeguards is punishable	X	—	—	—
If a license is not obtained, no insurance reimbursement for damage is allowed	—	—	X	—
If Captain, Master, or Engineer is drunk, racing, gambling, shall be wholly liable for death, injuries, and damage without any insurance	—	X	X (drunkenness not included)	X
If Captain, Master, or Engineer is drunk, racing, gambling, and there are deaths, injuries, or damage there is a mandatory sentence and fine	—	X	X (drunkenness not included)	X
Engineers will be tested for competence and licensed	—	—	X	—

Table 5. Comparing the Original Grundy Proposed Bill and the One Reported Out of Committee on January 9th

	December 6, 1837 Grundy's Proposed Law [44, p. 8; Appendix B]	January 9, 1838 Law Reported Out of Committee [46]
Technical Solutions		
Boiler and machinery inspections	X	X—only general testing retained
Certificate of boiler pressure limits	X	Portion of Section 5 Deleted
Inspectors shall test boiler pressures	X	Portion of Section 5 Deleted
Periodic testing shall take place	X	X—only general testing retained
Operational Solutions		
When boat is stopped, machinery will be kept in gear so that water pump is kept activated, and the safety valve must be kept open	X	Portion of Section 6 Deleted, and the requirement for an open safety valve is recommended when practical
Must shut off steam when passing ascending vessels at night	X	Section 10 Deleted

General Fire-fighting and Safety Solutions

Must carry long boats for safety	X	X
Must carry suction hose and fire hose	X	X
Night running requires signal lights	X	Modified in that the exact height above the bow deck was deleted

Civil Liability and Licensing Solutions

Licensing of all steamboats	X	X
If Captain, Master, or Engineer is drunk, racing, gambling shall be wholly liable for death, injuries, and damage without any insurance	X	X
If Captain, Master, or Engineer is drunk, racing, gambling, and there are deaths, injuries, or damage there is a mandatory sentence and fine	X	X

Some of the letters sent to the Committee during their deliberations (and still held in the National Archives) included:

1. A 12/18/1837 letter from A. B. Fannim of Georgia who asked for modifications in three technical sections of S.1.: Section 5 because of low pressure boats; in Section 7 because many boats could not detach the wheels from the engine and thus keep the pump going; and in Section 10 because some rivers were not wide enough.

2. A 12/23/1837 letter sent to Senator Nehimiah Knight from George Curtis who had submitted the S.1. draft to steamboat owner Robert Schuyler of New York and Captain Comstock who both asked for modifications in two technical sections, Section 5 on the pressure testing and Section 7 because many boats could not detach the wheels from the engine and thus keep the pump going.

3. A 1/8/1838 petition signed by five steamboat captains from Grundy's state of Tennessee who passed over the first four sections, but asked for modifications in Section 5 on the pressure testing because they felt that testing competence in the engineer working the boiler was more important and that the testing itself would weaken the boilers, and that most of the bursting problems occurred because the boiler was old or out of water. They also asked for modifications in Section 7 because many boats could not detach the wheels from the engine and thus keep the pump going, such boats would be at the mercy of the current, and a competent engineer would use the safety valve, open the fire doors, dampen the fires, and ensure a sufficiency of water; they questioned the need for the long-boats for abandoning ship in Section 8. Finally, they asked for changes in Section 10 because the ability to safely pass by is dependent on having power, not turning it off; and in Section 11 because placing the lights too high above the deck could blind the pilot.

Interestingly, what these three letters asked for from the Bill in the way of *deletions or amendments* did generally happen. Yet, when these outside experts asked for *additional* provisions to be included, such as the testing of a captain's or engineer's competence, they were not included.

In addition to the three letters listed above, the National Archives file for the Select Committee included a series of proposed additions for S.1. from Senator Nehemiah Knight that were sent in on 12/29/1837, and thus were available to the Committee before the Bill was reported out. This was the same senator who had received the George Curtis recommendations listed above on the 18th of December. The Knight proposal mainly comprised excellent and specific additions to Section 2 describing the kinds of inspectors that should be appointed along with their compensation and the specific things to be inspected, e.g., the safety valves, three water gauge cocks, mercury pressure gauges on the boilers, and an additional pump to pump water when the wheels are stopped. The Knight

Proposal also had a substantial list of the requirements for engineers and fireman as well as details on the management of engines and boilers [48].

None of these suggestions for additions was included in the Bill reported for a vote to the Senate [49].

It now appears that Grundy's Senate Select Committee took most of their bill from the previous Naval Affairs Committee Bill, then asked for slight modifications or deletions at the last minute arising from the correspondences from experienced captains and engineers, and refused any distinct additions to the technical solutions or the operational solutions. The bill that was reported out to the Senate was weighted heavily on the general safety and fire-fighting and civil liability solutions.

Perhaps another reason for the emphasis on the general safety and on civil liability solutions in S.1. was that Calhoun, who was on the Committee despite his distrust and fears of growing Federal power, was able to modify the law to keep the powers of the Federal government abstract and ill-defined; the more abstract, the less the possibility of Federal intrusions in state actions. Calhoun needed to be on the Committee to counter Waddy Thompson politically in South Carolina elections that Fall, but that did not mean that Calhoun had to support the legislation in its most specific aspects. In fact, the general safety and civil liability solutions were exactly like those proposed by the Charleston, South Carolina city government Committee of Twenty-one [19].

Interestingly, in the Senate debate on S.1. on January 23rd and 24th, most of the debate had to do with the civil liability solutions [47]. There was some discussion of general safety solutions to the problem, specifically requirements for life preservers and the placement of running lights. Senator Norvell, a Democrat from Michigan, even referred to the *Home* disaster in making his recommendations concerning life preservers. Other than that, the Senate debate as entered into the official record focused on what kind of evidence was *prima facie* in criminal proceedings for captains and engineers involved in steamboiler explosions. One amendment was passed that called for the presence of experienced engineers.

In a handwritten synopsis of S.1. prepared for the House of Representatives once the Senate had passed the bill [1], the meaning of some of the items was defined. For example, the focus of the inspections was to be on

- The age of the boat and boilers;
- When and where built;
- How long in use; and
- Whether the boat is, by the opinion of the inspector, fit to carry freight and passengers and the boiler fit for use.

The House primarily argued about the size of the fines when they considered the bill, and, when it was returned to the Senate for final passage and reconciliation, it was passed with little debate. In fact, as if to emphasize the focus upon the civil liability solutions, on July 5, 1838, when the final bill was passed by the Senate and sent to President Van Buren for his signature, Webster echoed what he had said some five years earlier in the Senate chamber when he had first proposed a bill. In 1833 he had said:

> But I look with more confidence of beneficial results from certain other provisions, which I trust will receive the consideration of the committee. Fully believing that these accidents generally result from negligence, at the time, by those who have the charge of the engine, penalties, I think, ought to be enacted against such negligence, and legal means provided, by which when lives are lost by such occurrences, an immediate inquisition, investigation, and trial should be secured, and the culpable negligence, if there be such, adequately published [11, pp. 765-766; 12, p. 647].

In 1838 he said in quite a similar fashion:

> What is the remedy for such negligence? It may be looked for to the owners of steamboats, and no doubt they were individually responsible, in a civil point of view, for every loss by such disasters. It would fall on the owners of the boats by the principles of the common law [50, p. 2].

In five years, the basis of solving the steamboat explosion problem had changed very little save to add the side considerations of safety and fire-fighting.

Some had little hope that the law would in fact work to prevent boiler explosions. The Charleston City Council in their report said as much when they too largely ignored the technical and operational solutions. The engine and its lack of controls and gauges were fine—it was usually the fault of the engineer alone:

> The Engine will do its duty faithfully, if the Engineer will do his. In this *particular*, it is to his vigilance and skill that the traveler must look for security; for in his hands while the vessel moves (without meaning to be irreverent) may it be said, are the "issues of life and death." A single moment of inattention, a small indiscretion, a rash and heedless self-will, may not only cost him his own life, but hurry to destruction the lives of his fellow-creatures. Can human laws, or human ingenuity guard against the abuse of the tremendous power entrusted to this person? We fear not; for even with the punishment of death staring him in the face, the veteran and tried soldier has been known to sleep upon his post [19, p. 30].

The 1838 report from the scientists at the American Institute in New York City concluded in a similar fashion:

> The act passed at the last session of Congress, "to provide for the better security of the lives of passengers on board steam-vessels," although not what your committee suppose to be the best that could have been devised, is still, in their opinion, likely to be productive of much good, if those are employed, in carrying it into effect, possess sufficient skill and integrity to fit them to the task [51, p. 651].

Endnotes

1. Papers of the Select Committee on Bill S.1, 25th Congress, Session 2, National Archives, Legislative Division.
2. *National Intelligencier, 25*:7572, p. 3, May 19, 1837. See also *25*:7575, p. 3, May 23, 1837; *25*:7576, p. 3.
3. *Niles Register, 52*, March 18, 1837.
4. Major L. Wilson, *The Presidency of Martin Van Buren,* University of Kansas Press, Lawrence, Kansas, 1984.
5. Donald B. Cole, *Martin Van Buren and the American Political System*, Princeton University Press, Princeton, New Jersey, 1984.
6. Merrill D. Peterson, *The Great Triumvirate: Webster, Clay, and Calhoun*, Oxford University Press, New York, 1987.
7. Read on 12/4/1837. *Journal of the Senate*, 25th Congress, 2nd Session.
8. Robert V. Remini, *Daniel Webster, The Man and His Time*, W. W. Norton, New York, 1997.
9. Robert V. Remini, *Andrew Jackson and the Course of American Democracy, 1833–1845*, Harper & Row, New York, 1984.
10. *Sketches of United States Senators, of the Session of 1837–8*, William M. Morrison, Washington, D.C., 1839.
11. *The Papers of Daniel Webster, Legal Papers, Volume 3: The Federal Practice, Part II*, Andrew J. King, Editor, University Press of New England, Hanover, New Hampshire, 1989.
12. The General Improvement of Society, *Quarterly Christian Spectator, 6*, December 1834.
13. Michael Birkner, *Samuel L. Southard: Jeffersonian Whig*, Associated University Presses, Cranbury, New Jersey, 1984.
14. Thomas Benton,*Thirty Years' View*, D. Appleton, New York, 1856: I. At one point Calhoun had such enmity for Van Buren that when he knew his vote would prevent Van Buren's nomination as minister to Great Britain in 1831, Calhoun supposedly exclaimed: "It will kill him, sir, kill him dead."
15. S. A. Howland, *Steamboat Disasters and Railroad Accidents in the United States*, Dorr, Howland & Co., Worcester, 1846.
16. John H. Morrison, *History of American Steam Navigation*, Stephen Daye Press, New York, 1958.
17. The Rev. Thomas Smyth, *The Voice of God in Calamity: or Reflections on the Loss of the Steam-Boat Home, October 9, 1837. A Sermon Delivered in the Second Presbyterian Church, Charleston on Sabbath Morning, October 22, 1837*, Jenkins & Hussey, Charleston, South Carolina, 1837.
18. *Charleston Mercury, 26*:4317, p. 3, October 23, 1837; *26*:4319, p. 3, October 25, 1837; *26*:4334, p. 3, November 13, 1837.
19. *The Proceedings of the Citizens and City Council of Charleston, in Relation to the Destruction of the Steamboat Home*, Thomas Eckles, Charleston, 1837.
20. William C. Redfield, Correspondence with United States Board of Navy Commissioners, *Journal of the Franklin Institute, 12* (3rd Series), July 1846. A revisionist critique of the *Home* disaster.
21. *The Papers of Daniel Webster, Correspondence, Legal Papers, Vol. 3. The Federal Practice*, Andrew J. King, Editor, University Press of New England, Hanover, New Hampshire, 1989.

22. *National Intelligencier*, p. 2, col. 3, 6/25/1838.

23. The actual only record we now have of Calhoun directly receiving this petition from Charleston was on July 2, 1838.

24. Letter of Felix Grundy, 1/18/1838, Southern Historical Collection, University of North Carolina, Chapel Hill. "The hatred of these men [the abolitionists] to Mr Van Buren is unbounded, merely because, he had pledged himself, in his inaugural address, to veto any measure, which may have a tendency to accomplish their object—This, in their eyes, is an offense, not to be pardoned—They know very well, he will never abandon the ground he has taken on this subject; he therefore presents an obstacle to their wishes, which can only be removed, by his prostration—In this, I trust they will be disappointed, because, I verily believe, Mr Van Buren from his northern position [Van Buren did come from New York], in addition to other reasons, can do more than any other man to put down this dangerous and disorganizing spirit."

25. Irving H. Bartlett, *John C. Calhoun: A Biography*, W. W. Norton, New York, 1993.

26. Henry Tazewell Thompson, *General Waddy Thompson: Member of Congress, 1835-41, Minister to Mexico, 1842-44*, E. Bridges, Greenville, South Carolina, 1925 (reprinted).

27. Letter of James Edward Calhoun, April 21, 1838 (quoted in [26]).

28. Attacks on Legaré's vote against the Administration's sub-treasury appeared in the *Charleston Courier* on April 24, 1838 and May 17th, 1838 signed anonymously by "A Constitutent." Legaré was defeated that Fall, but Thompson won re-election handily [26, pp. 6-7].

29. James L. Pettigru to Thompson within a month of the Fall election results: "If you could have imparted something of your resolution to our friend Legaré he would not have been the sport of popular caprice that he is now. . . . The 'holy alliance' between Poinsett and Calhoun has regulated without difficulty the Public of Charleston." (quoted in [26, p. 8]).

30. Joseph Howard Parks, *Felix Grundy: Champion of Democracy*, Louisiana State University Press, Baton Rouge, 1940.

31. Samuel Rexford Mitchell, *The Congressional Career of Felix Grundy*, A Dissertation Submitted to the Faculty of the Department of History, University of Chicago, Master of Arts, September 1925.

32. Elbert B. Smith, *Magnificent Missourian: The Life of Thomas Hart Benton*, J. B. Lippincott Co., Philadelphia, 1957.

33. Alexander Mackay, *The Western World; or Travels in the United States in 1846–47*, (3 vols.), R. Bentley, London, 1849.

34. Richard Arden Wire, *John M. Clayton and the Search for Order; A Study in Whig Politics and Diplomacy*, PhD Dissertation, University of Maryland, 1971 (University Microfilms, Ann Arbor Michigan–71-25,973). Thomas earlier filled out the term of Caesar Rodney, and was re-elected in 1841 after filling out John's later unexpired term. When John returned to the Senate in 1844, Thomas resigned his seat in the Senate, and died four years later.

35. Donald B. Cole, *Martin Van Buren and the American Political System*, Princeton University Press, Princeton, New Jersey, 1984.

36. Edwin Arthur Miles, *Robert J. Walker—His Mississippi Years*, a Masters of Arts in History thesis, University of North Carolina, Chapel Hill, 1949.

37. Peter J. Parish, Daniel Webster, New England, and the West, *Journal of American History, 54,* December 1967.
38. Calhoun also had used the Senate select committee maneuver in 1835 in order to offer legislation on "incendiary" abolitionist literature (quoted in [31], p. 74).
39. James P. Shenton, *Robert John Walker: A Politician from Jackson to Lincoln,* Columbia University Press, New York, 1961.
40. *Report of the Committee on Commerce, Accompanied by a Bill for Regulating of Steam Boats, and for the Security of Passengers Therein,* 18th Congress, Session 1, House Reports, No. 125, Gales & Seaton, Washington, 1824 (Serial Set 106).
41. Report No. 478, Steamboats, May 18, 1832, in *Reports of Committees of the House of Representatives at the First Session of the Twenty-Second Congress, Begun and Held at the City of Washington, December 7, 1831,* Duff Green, Washington, 1831.
42. Derived from *The Congressional Globe, 1*:4, December 23, 1833.
43. Derived from *Journal of the Franklin Institute, 14*:4, October 1834.
44. Derived from original proposed Bill as printed in *The Congressional Globe, 6*:1, December 11, 1837.
45. *Correspondence of Andrew Jackson,* John S. Bassett, Editor, Carnegie Institute, Washington, D.C., 1931.
46. John K. Brown, *Limbs on the Levee: Steamboat Explosions and the Origins of Federal Public Welfare Regulation, 1817–1852,* International Steamboat Society, Middlebourne, West Virginia, 1989.
47. *Journal of the Senate, 25th Congress, 3rd Session,* Blair and Rives, Washington, D.C., 1838.
48. "Document in relation to a mode of preventing the explosions of steam boilers. 1837, Dec. 29 referred to the Select Committee to whom was referred the Bill S.1. 25th Congress, 2nd Session, p. 5." National Archives, Center for Legislative Archives, SEN25A-B1—B6, D19 Boxes 1, 4, 6, 8 and 27.
49. *National Intelligencier, 26*:7770, January 8, 1838.
50. National Intelligencer, 26:7924, column 2, July 7, 1838.
51. John D. Ward, T. B. Stillman, and Stephen Dod, Report on the Causes, &c, of the Explosion of Steam-Boilers, *Journal of the American Institute, III*:12, September 1838. The American Institute in New York City, a sister institution to the Franklin Institute, also began investigating steam-boiler explosions in July 1837 and published their findings in September 1838.

Figure 29. *Steamboat Bill* [1-3].

Down the Mississippi steamed the Whipperwill,
Commanded by that pilot Steamboat Bill.
The owners gave him orders on the strict Q. T.—
to try and beat the record of the Robert E. Lee.
Just feed up your fires, let the old smoke roll.
Burn up all your cargo if you run out of coal.
If we don't beat that record, Billy told the mate,
Send my mail in care of Peter to the Golden Gate.

Steamboat Bill, streaming down the Mississippi,
Steamboat Bill, a mighty man was he,
Steamboat Bill, steaming down the Mississippi,
Going to beat the record of the Robert E. Lee.

Up then stepped a gambling man from Louisville,
Who tried to get a bet against the Whipperwill,
Billy flashed a roll that surely was a bear,
The boiler it exploded, blew them up in the air.
The gambler said to Billy as they left the wreck,
I don't know where we're going, but we're neck n' neck.
Says Billy to the gambler, I'll tell you what I'll do,
I will bet another thousand I'll go higher than you.

Steamboat Bill, he tore up the Mississippi,
Steamboat Bill, the tide it made him swear,
Steamboat Bill, he tore up the Mississippi,
The explosion of the boiler got him up in the air.

River's all in mourning now for Steamboat Bill,
No more you'll hear the puffing of the Whipperwill.
There's crape on every steamboat that plows those streams
From Memphis right to Natchez down to New Orleans.
The wife of Mister William was at home in bed,
When she got the telegram that Steamboat's dead.
Says she to the children,
"Bless each honey lamb, the next papa that you have will be a railroad man.

Steamboat Bill, missing on the Mississippi,
Steamboat Bill, is with an angel band,
Steamboat Bill, missing on the Mississippi,
He's a pilot on a ferry in that Promised Land.

CHAPTER 5

The Law Didn't Work

... the news was recd. of the loss of the *Pulaski* steamer (see Figure 30), on her passage from Charleston to Baltimore, & of more than 200 persons on board her only 15 saved! Her loss occasioned by the bursting of her boiler, which caused her to sink. The Southern members, who had friends on board her look sad and solemn. Gov. Hamilton [South Carolina] & his two sons, it is said were on board her. Such losses are dreadful. The Nation must shake off its apathy about steam navigation and arouse itself; legislation must be had on the subject or nobody's life will be safe who travels. Only last October the Home was lost with many valuable lives—recently the *Moselle* was blown up & hundreds of human beings hurried into eternity in a moment—the paper of today contains the account of the burning of the *Washington* on Lake Erie & the loss of more than 50 lives! Something must be done, or steam navigation had better never been discovered—it will prove, to the human race rather a curse than a blessing [4, p. 88].

The steamboat bill of 1838 did not work [5, 6]. Many of the parlor songs about steamboat disasters included at the beginnings of chapters in this book, came **after** the passage of the 1838 S.1. Law. To show the utter worthlessness of the inspections, consider the following testimony given by an official steamboat inspector at a coroner's inquest where several lives were lost in 1846 on the *Lexington* [7-9]:

Inspector John Clark: "I only consider myself bound to inspect the hull, boiler, and machinery; I go first on deck, then below, and look about, when I inspect a vessel—nothing more than look at the wood and iron."

Coroner: "Have you ever condemned a boat?"

Inspector Clark: "We never condemned any boat. We have restricted them to a certain amount of steam."

Juror: "When you inspect a boat you look at the wood and do nothing else?"

Inspector Clark: "Yes, we take our fees."

Juror: "How do you examine the hull of a vessel?"

Inspector Clark: "Why, I examine it."

Figure 30. The explosion of the steamboat *Pulaski* after
the passage of S.1.

Juror: "How?"

Inspector Clark: "With my eyes."

Juror: "Well, we want to know your mode of proceeding."

Inspector Clark: Well, I go and inquire the boat's age;—How much do you
suppose I am to do for five dollars?"

Juror: "No matter, sir, about the fee; we want to know your mode of
proceeding."

Inspector Clark: "Why, I examine the hull, and I look at the engine; I have
worked at and made almost every kind of engine for the last thirty years."

[Here, the coroner stated he had some doubts about the propriety of examining the witness as to particular steamboats, except as they referred to the
case of the *Lexington*.]

Juror: "In due deference to your better judgement, sir, I think he is bound to
say how he proceeds in the examining of steamboats. It is the opinion of
the jury, that the inspectors have passed steamboats as safe, and given
them certificates when they are not worthy of it" [10, pp. 24-25].

It wasn't until Burke, the Commissioner of Patents, re-published the Franklin
reports in a combined fashion in 1848 that they were considered to be "the present
state of knowledge in relation to this subject." Perhaps it took time for the Franklin

Reports to become generally accepted, and time for the Federal Government to recognize that an emphasis on civil liability solutions at the price of specific technical solutions would not solve the problem. The emphasis on civil liability grew out of the contemporary view on technology that saw the whole steam engine system under the conscious control of the engineers and captains who would only overload the engines through neglect, carelessness, or mis-use during racing.

The idea that "racing" was at the root of the problem held sway in the public imagination for many years as evidenced by its use as a plot device in many of the most popular novels of the day:

- While *The Scarlet Letter* sold only 10,000 copies during Hawthorne's life-time, *The Lamplighter* sold 40,000 in its first month in 1854 [12, p. ix], and in Chapter 41, a steamboat explosion and fire while racing was a crucial plot element [12, pp. 324-330].
- *Clotel; or the President's Daughter* was, arguably, the first African-American novel in 1853 [13, p. x], and Chapter 2 used a steamboat explosion resulting from a race, even noting how the engineer secured the safety valve and permitted no steam to escape [13, p. 58; 14].

Perhaps it took time for gauges to be generally installed on all steamboats as the early generation of engines finally, after repeated salvagings, retired.

It took time for small auxiliary pumps called "doctors" to be installed on a large number of steamboats, and it was these doctors that would pump water into the boiler when the large main engine was stopped [15, p. 162]—the final effective solution to the X-ed out material in the Grundy's draft law (see Figure 1).

It certainly took time for technological mastery to be developed in engineers and firemen and for apprentices to be properly trained.

Moreover, perhaps it took time for Congress, the States, the Judiciary, and the President to get used to the idea of interstate commerce regulation—S.1. was the first example of such legislation in the United States.

A new law was passed by the Federal Government in 1852 that enacted many of the provisions suggested in the debate around S.1. [16, p. 78]. Beginning in 1852,

- all boilers would be pressure tested on a regular basis;
- all engineers, pilots, and captains would be tested and licensed (see Figure 31);
- all boilers would have fusible plugs that would blow out before dangerously high levels of boiler pressure were reached;
- effective new gauges and indicators would have to be installed; and
- a powerful group of independent inspectors could board any vessel at any time and require repairs as needed by their inspections.

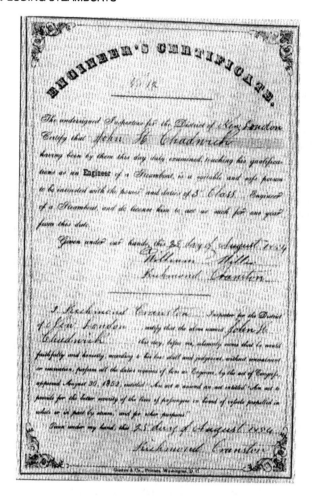

Figure 31. Steam engineer's certificate from new 1852
steamboat law [18, p. 153].

This new law worked; the year before it took effect there were 1,038 fatalities
and, in the year following, 45. In the eight years that followed the passage of the
law, deaths from boiler explosions fell by thirty-three percent [17, p. 71].

The problems with passing an effective steamboat law in 1838 arose from
many different sources including:

• A early technology without effective methods of control, without a history of
  human interaction to base training, and with problems liable to be hidden and
  complex.

Figure 32. The figure of Kairos balancing the particulars on a scale
while himself balancing on the edge [19, p. 32].

- Lurid newspaper reports and illustrations, parlor songs, and tall tales causing the public to overreact to the steamboat explosions and to fix on human incompetence rather than systemic technological problems.
- A misunderstanding of the evolution of the steamboat in which the part apparent to all—the hulls—changed quickly and dramatically, but the less apparent internal engines boilers and controls—changed much more slowly.
- Non-technically oriented Congressional readers suffering from information overload. The number and size of the reports on the desks of the Congress in 1838 were overwhelming, and the reports, all save Bache's *General Report*, were not designed to aid a reader's understanding. They were intentionally kept unstructured, and this lack of structure in **large** reports made the congressmen vulnerable to over-representing **short** letter reports landing on their desks at the end of the whole process.
- Senators and Congressmen who with a nearly universal legal background sought solutions in familiar areas, in areas of civil liability.
- A Federal government that had never passed an interstate commerce regulation and was unsure as to its constitutionality especially in the atmosphere of nullification attacks on the integrity of Federal power.

- Elements of the regulation that came to be identified too closely with one party, the Whigs and Webster, and thus would invoke the ire of the opposing Democrats and the Calhoun nullifiers.
- Political expediency by a weakened president that both motivated and permitted those who were innately opposed to an effective law, Senator John C. Calhoun, to participate in its devising. Once included in the Select Committee and without an effective check on his political power, Calhoun was able to disable the enforcement of the law's specific effects.

Yet despite all these problems, an effective law was finally passed in 1852 because 1852 was the moment of *kairos*.

The ancient Greeks had a sense of time as both the chronological passage of time—chronos—just as the antebellum explosions and disasters in this story chronologically progressed from year to year and month to month. However, they also had a concept called kairos (see Figure 32), the "exact or critical time, season or opportunity" [19, p. 31]. Kairos was a sense of multidimensional "ripeness" of time, a moment when, so to speak, a message can be heard and carried into action. As Isocrates wrote:

> The moment for action has not yet gone by, and so made it now futile to bring up this question; for then, and only then, should we cease to speak, when the conditions have come to an end and there is no longer any need to deliberate about them [19, p. 33].

Evidently, the moment of kairos had not yet arrived in 1837 and 1838 for the words of Bache and the other Franklin Institute scientists to reach the ears of Grundy, Calhoun, and the Congress. The pages of Bache's report preceded the necessary technical innovations by just too much. The lurid adjectives of the newspaper accounts and the melancholic songs had overwhelmed decisive rational actions, and the ability of Congress and the people to do anything save wish in vain that steam navigation had better never have been discovered.

Yet look again at the words of the parlor song beginning this Part, *Steamboat Bill*. The music created by the Leighton Brothers in 1910 was quite lively, and had more of a likeness to a ragtime tune than any of the prior songs. They were more like a dirge, in minor keys, and designed to prick the emotions of melancholy. *Steamboat Bill*, on the other hand, was a dance or a comedic tall-tale to sing. Also, the words by Shields contain none of the pathos of the earlier songs, and the effect of having the boilers blow is just a device to get the Bill and the gambler blown into the air for another bet. There are no dead bodies, no mourning; why even Bill's widow is already plotting what type of husband she'll have next.

The song *Steamboat Bill* as early as 1910 signaled that American culture had come to terms with the ambivalence surrounding the steamboat technology. The technology had been so safely harnessed that by the time of Robert Grimshaw's

1885 *Steam Engine Catechism*, there are no items in the index related to boiler explosions or accidents. So the sense that humans were playing with the devil in the pressures of a steamboat was gone. Moreover, the sense of the steamboat as a gift from the gods in its crucial role in western transportation was progressively eclipsed by the railroads. The whole problem of boiler explosions no longer haunted the imagination, and, with the passage of time, the dirge and deaths turned into the raw material for humor as can be seen in the 1910 song or in the 1928 silent movie by Buster Keaton, *Steamboat Bill Jr.* [21], or it moved in the life of Mark Twain [20] from a point of horror in regards to his brother's death on the exploding *Pennsylvania* in 1858, to a neutral metaphor in his 1889 *A Connecticut Yankee in King Arthur's Court*:

> When they were within fifteen yards, I sent that bomb with a sure aim, and it struck the ground just under the horses' noses.
>
> Yes, it was a neat thing, a very neat and pretty to see. It resembled a steamboat explosion on the Mississippi; and during the next fifteen minutes we stood under a steady drizzle of microscopic fragments of knights and hardware and horse-flesh [22, p. 262; 23].

In a way, many new technologies at the moment of entering a culture were seen in a similar ambivalent position:

- Electricity lit up the Chicago 1880 exposition and electric girls covered with bulbs worked as maids [24, p. 208], yet electricity was the reason Frankenstein existed, and Westinghouse repeatedly pointed out how Edison's AC electricity was used to power prison electric chairs at the turn of the century;
- Early electric automobiles had a great liberating effect on women:

> Despite the traditional association of the automobile as a mechanical object with men and masculinity in American culture, automobility probably has had a greater impact on women's roles than on men's. . . . Because driving an automobile requires skill rather than physical strength, women could control one far easier than they could a spirited team. They were at first primarily users of electric cars, which were silent, odorless, and free of the problems of hand-cranking to start the engine and shifting gears. Introduction of the self-starter in 1912, called the "ladies' aid," and of the closed car after 1919, which obviated wearing special clothes while motoring, put middle-class women drivers in conventional gasoline automobiles in droves [25, p. 162].

And yet the public also saw them as the reason why 1920s gangsters and gunmen could speed across state lines and elude the police. Clyde Barrow of Bonnie and Clyde even wrote the following letter to Henry Ford describing the usefulness of the automobile in his life of crime:

Mr. Henry Ford
Detroit Michigan
Dear Sir:—
While I still have got breath in my lungs I will tell you what a dandy car you make. I have driven Fords exclusively when I could get away with one. For sustained speed and freedom from trouble, the Ford has got every other car skinned, and even if my business hasn't been strictly legal, it don't hurt any thing to tell you what a fine car you got in the V8.
Your's truly,
Clyde Barrow [26]

Computers, atomic energy, planes, all technologies affect culture in an ambivalent manner at the moment of their first appearance in a culture . . . and steamboats in the 1820s and 30s were no different.

Endnotes

1. Words by Ben Shields, Music by Leighton Bros, F. A. Mills, New York, 1910. Based on the life of William L. Heckman, a Missouri River pilot who wrote stories under the pseudonym of Steamboat Bill.
2. Captain William L. Heckman, *Steamboating: Sixty-five Years on Missouri's Rivers— The Historical Story of Developing the Waterway Traffic on the Rivers of the Midwest*, Burton Publishing Company, Kansas City, 1950.
3. Breckinridge Collection in the Western Historical Manuscript Collection— Columbia, University of Missouri/State Historical Society of Missouri, for a collection of newspaper pieces written by Heckman entitled "Steamboating in the Old Days."
4. June 21, 1838 journal entry from Benjamin B. French, *Witness to the Young Republic: A Yankee's Journal, 1828-1870*, University Press of New England, Hanover, New Hampshire, 1989. In 1838 French was an assistant clerk in the House of Representatives.
5. *Memorial of Sundry Proprietors and Managers of American Steam Vessels on the Impolicy and Injustice of Certain Enactments Contained in the Law Relating to Steamboats*, New York, 1840.
6. *Proceedings of a Meeting and Report of a Committee of the Citizens of Cleveland in Relation to Steamboat Disasters on the Western Lakes*, Steam Press of Harris, Fairbanks & Co., Cleveland, Ohio, p. 7, 1850. See also p. 16: "The law now in existence, passed July 7, 1838, entitled 'An Act to provide for the better security of the lives of passengers on board of vessels propelled in whole or in part by steam,' has many valuable provisions in it, but it is defective in several essentials."
7. In this same case of the sinking of the *Lexington*, Daniel Webster was again deeply involved [8; 9, pp. 789-790]. He was exactly where he was when he argued against the monopoly of Fulton in 1824, he was back as a lawyer in the Supreme Court. This time he was representing a suit brought against the owners of the steamboat *Lexington* for negligence.

8. Rev. E. H. Chapin, *A Discourse on the Burning of the Steamboat Lexington: Preached in the First Independent Christian Church, March 8th and 15th,* James C. Walker, Richmond, 1840.

9. *The Papers of Daniel Webster, Legal Papers, Volume 3: The Federal Practice, Part II,* Andrew J. King, Editor, University Press of New England, Hanover, New Hampshire, 1989.

10. Senate # 241, 25th Congress, 1st Session: pp. 22-24; reprinted in [11, p. 591].

11. John H. Morrison, *History of American Steam Navigation,* Stephen Daye Press, New York, 1958.

12. Maria Susanna Cummins, *The Lamplighter,* edited and introduced by Nina Baym, Rutgers University Press, New Brunswick, New Jersey, 1988.

13. William Wells Brown, *Clotel; or the President's Daughter,* in *Three Classic African-American Novels,* Edited and with an introduction by Henry Louis Gates, Vintage Books, New York, 1990.

14. Susan Warner, *The Wide, Wide World,* Afterword by Jane Tompkins, The Feminist Press at the City University of New York, New York, 1987. This book is somewhat an exception to 12 and 13 in that when the heroine Ellen is aboard a steamboat, in Chapter 7 and 8, it does not blow up, and, in fact, friendly words from a stranger aboard the steamboat aid her in staying on the "straight and narrow."

15. Louis C. Hunter, *Steamboats on the Western Rivers: An Economic and Technological History,* Harvard University Press, Cambridge, 1949.

16. Robert F. Bennett, A Case of Calculated Mischief, *U.S. Naval Institute Proceedings, 102*:3, March 1976.

17. John K. Brown, *Limbs on the Levee: Steamboat Explosions and the Origins of Federal Public Welfare Regulation, 1817–1852,* International Steamboat Society, Middlebourne, West Virginia, 1989.

18. Douglas Stein, *American Maritime Documents: 1776–1860,* Mystic Seaport Museum, Mystic, Connecticut, 1992.

19. Sharon Crowley and Debra Hawhee, *Ancient Rhetorics for Contemporary Students,* Allyn and Bacon, Boston, 1999.

20. Justin Kaplan, *Mark Twain and His World,* Simon & Schuster, New York, 1974.

21. 1928 silent movie, *Steamboat Bill, Jr.* New York, NY. Story by Carl Harbaugh. Photography by Dev Jennings and Bart Haines. Cast included: Buster Keaton, Marion Byron, Tom Lewis, and Ernest Torrence.

22. Mark Twain, *A Connecticut Yankee in King Arthur's Court,* Harper & Brothers, New York, p. 262, 1889. This movement from sheer horror in regards to his brother's death on the *Pennsylvania* in 1858 through his passionate retelling of the incident in Chapter 4 of the *Gilded Age* in 1873 and then in *Life on the Mississippi* in 1883 to appear as a relatively neutral device used by Huck in Chapter 32 of 1885's *Huckleberry Finn* (Now I struck an idea, and fetched it out: "It wasn't the grounding—that didn't keep us back but a little. We blowed out a cylinder-head." "Good gracious! Anybody hurt?" [asks Aunt Sally] "No'm. Killed a nigger.") to finally a neutral metaphor in 1889's *Connecticut Yankee.* A three decade move from horror to metaphor in Twain's literary consciousness mirrored the change in consciousness of the American public in regards to steamboat explosions.

23. Edgar Branch, *Men Call Me Lucky: Mark Twain and the Pennsylvania,* Friends of the Library Society, Miami University, Miami, Ohio, 1985.

24. Carolyn Marvin, Dazzling the Multitude: Imagining the Electric Light as a Communications Medium, in *Imagining Tomorrow: History, Technology, and the American Future*, Joseph J. Corn, Editor, The MIT Press, Cambridge, Massachusetts, 1986.
25. James J. Flink, *The Automobile Age*, MIT Press, Cambridge, Massachusetts, 1988.
26. On display in Greenfield Village museum, Detroit, Michigan.

Glossary

Atmospheric steam engines—see low-pressure steam engines.

Cast iron—an early process of producing iron in which the iron was cast in a mold and was hard, brittle, and nonmalleable in comparison to wrought iron.

Damping—closing the dampers, the air ducts to the boiler fire, and thus effectively turning down the fire.

Doctor—an auxiliary engine that would supply water to the boiler separate from an engine used for propulsion.

Expunging bill—Legislative action to expunge from the Congressional record the Whig's reprimand of President Andrew Jackson's imperious gestures in an earlier Congress.

Foaming—Bubbles created by the boiling of the water in a boiler which would cause water to issue from a stop-cock—seeming to indicate that there was water at or above the stop-cock—even though it was foam carrying the water up to the stop-cock from several inches below.

Freeboard—the distance from the waterline on the hull to the top of the deck.

Fusible plates (or plugs)—a safety device allowing pressure above a specified level to intentionally blow out of a boiler. These plates or plugs would blow out rather than the whole boiler because they were made of an alloy which melted at lower temperatures than did cast or wrought iron.

High-pressure steam power—a steam engine in the boiler would create high pressure to push a piston out, and in comparison with low-pressure engines used less fuel and required less weight in the engine.

Kairos—an exact or critical time, season, or opportunity in contrast to chronos in which each moment of time is homogeneous.

Low-pressure steam engine—the earliest steam engine invented by Watt and Newcommen in which steam, during its condensation phase, would shrink in mass and develop a vacuum that would pull a piston. These engines were termed "low pressure" or "atmospheric" steam engines, and, for their large size and heavy weight, could only develop low levels of power. Moreover, these "low-pressure" engines required a vast amount of fuel and cold water to produce a low level of power.

Nullification/Nullifier—the alleged right of a state to suspend operation of a federal law within its boundaries. Nullification was based on a belief that states were the ultimate sources of sovereignty, and that the federal government was simply a league of freely associated states, the authority of which the state was free to recognize or ignore in accordance with its best interests. The specific situation giving rise to Calhoun and South Carolina's nullification efforts was a high protective tariff which impeded Carolina's cotton trade with England.

Pre-emption—legislative initiatives to make such western lands free and open to squatter settlers.

Safety barge—a barge towed by lines behind a steamboat and on which there would be no engine and thus no possibility of explosion. Used primarily on the Hudson River.

Safety valves—a safety device which would allow excess pressure in a boiler to blow off harmlessly rather than explosively. It was basically a lever and fulcrum with a weight attached to a rod at one end that kept a valve shut at the other end until the pressure of the steam exceeded the weight on the rod at which time the valve would open and the steam above the weight on the rod would be expelled.

Stop-cocks (a.k.a. gauge-cocks)—a safety apparatus in which two spigots on the side of a boiler, one above the other by some few inches, would directly measure the level of the water and steam in the boiler because water or steam issuing from them when they were opened would indicate the level of steam or water inside the boiler—e.g., if water issued, then the valve was below or at the water line; if only steam issued, then the water was below that stop-cock. This apparatus was confounded by the problem of foaming.

Wrought iron—a later development of iron production that produced an iron that was tough, malleable, and relatively soft in comparison to cast iron.

APPENDIX 1

Comparing the Four Legislative Attempts

The four attempts at steamboat legislation can be summarized as shown in Table 6. Moving from left to right, from Vinton's bill to that of the Committee on Naval Affairs, three general trends seem evident. In all 19 different regulations proposed in the four bills, there were some areas of consensus: three regulations were repeated in all of the bills, one was repeated in three of the bills, and nine were repeated in at least two legislative attempts. Probably such overlap in the bills was a function of each bill refashioning elements of prior bills. For example, Wickliffe's House proposals regarding operational and general fire safety were almost repeated verbatim in the Naval Affairs bill. Consider the two Section 7's below; only 10 words are changed out of 186.

Wickliffe's 1832 House Proposal

Sec. 7. And it be further enacted, That whenever the master of any boat or vessel, or the person or persons charged with the navigating said boat or vessel, which is propelled in whole or in part by steam, shall stop the motion or headway of said boat or vessel, or when the said boat or vessel shall be stopped for the purpose of discharging or taking in cargo or passengers, or when "wooding," and the steam in said boiler shall be equal to one ————, the ascertained strength of said boiler, he or they shall keep the engine of said boat or vessel in motion sufficient to work the pump, give the necessary supply of water, and to keep the steam down in said boiler to what it is when the said boat is underway, at the same time lessening the weight upon the safety-valve, so that it shall give way when the steam in said boiler is equal to one ———— of its ascertained strength, under the penalty of two hundred dollars for each and every offense.

Naval Affairs Committee's 1834 Senate Proposal

Sec. 7. And it be further enacted, That whenever the master of any boat or vessel, or the person or persons charged with the navigating said boat or vessel, which is propelled in whole or in part by steam, shall stop the motion or headway of said boat or vessel, or when the said boat or vessel shall be stopped for the purpose of discharging or taking in cargo or passengers, or when

Table 6. Comparing the First Four Steamboat Bills Proposed

	1824 Vinton's Proposed Law [1]	1830-2 Wickliffe's Proposed Law [2]	1833-4 Webster's Proposed Law [3]	1834 Committee on Naval Affairs' Proposed Law [4]
Technical Solutions				
Boiler and machinery inspections	X	X	X	X
Certificate of boiler pressure limits	X	X	X	X
Two safety-valves in place	X	—	—	—
One of two safety-valves locked	X	—	X	—
Pressure gauge included	—	—	X	—
Inspectors shall test boiler pressures	X	—	X	X
Periodic testing shall take place	—	X	—	X
Operational Solutions				
When boat is stopped, machinery will be kept in gear so that water pump is kept activated, and the safety valve must be kept open	—	X	—	X
Must shut off steam when passing ascending vessels at night	—	X	—	X

General Fire-fighting and Safety Solutions

Item	1	2	3	4
Must carry long boats for safety	X	—	X	—
Must carry suction hose and fire hose	X	—	X	—
Night running requires signal lights	X	—	X	—

Civil Liability and Licensing Solutions

Item	1	2	3	4
Licensing of all steamboats	X	X	X	X
Transportation of gunpowder aboard ship requires special placement and handling	X (a singular item in reaction most probably to Lioness disaster)	—	—	—
Tampering with Safeguards is Punishable	—	—	X	—
If a license is not obtained, no insurance reimbursement for damage is allowed	X	—	—	—
If Captain, Master, or Engineer is drunk, racing, gambling, shall be wholly liable for death, injuries, and damage without any insurance	X (drunkenness not included)	X	—	—
If Captain, Master, or Engineer is drunk, racing, gambling, and there are deaths, injuries, or damage there is a mandatory sentence and fine	X (drunkenness not included)	X	—	—
Engineers will be tested for competence and licensed	X	—	—	—

"wooding," and the steam in said boiler shall be equal to one **third** of the ascertained strength of said boiler **or boilers**, he or they shall keep the engine of said boat or vessel in motion sufficient to work the pump, give the necessary supply of water, and to keep the steam down in said boiler to what it is when the said boat is underway, at the same time lessening the weight upon the safety-valve, so that it shall give way when the steam in said boiler is equal to one ————— of its ascertained strength, under the penalty of ————— dollars for each and every offense **in neglecting or violating the requirements of this section**.

Endnotes

1. Derived from *Report of the Committee on Commerce, Accompanied by a Bill for Regulating of Steam Boats, and for the Security of Passengers Therein*, 18th Congress, Session 1, House Reports, No. 125, Gales & Seaton, Washington, 1824 (Serial Set 106).
2. Derived from Report No. 478, Steamboats, May 18, 1832, in *Reports of Committees of the House of Representatives at the First Session of the Twenty-Second Congress, Begun and Held at the City of Washington, December 7, 1831*, Duff Green, Washington, 1831.
3. Derived from *The Congressional Globe, 1*:4, December 23, 1833.
4. Derived from *Journal of the Franklin Institute, 14*:4, pp. 217-222, October 1834.

Index

William H. Morton, a safety barge, 88
Wilson, Majorie L., 16, 38, 117
Wire, Richard Arden, 118

Woodbury, Levi (Secretary of the
Treasury), 88
Wrought iron, 69, 82, 133